Pteranodon

-240	-220	-200	-180	-160	-140

-201.3 -145

Triassic   Jurassic

Pteranodon was a pterosa
It is one of the most well-
Pteranodon actually flew in
wings instead of flapping i
to fly freely in the sky w
"Toothless Wing." It is sup
preyed on fish.
●About 20-30 feet or 6-9 m

Three-Digit Addition

Level ☆
Score

/100

Math
DAY
1

Date   /   /

Name

1 Add.

5 points per question

(1)
```
  1 0 0
+   5 5
```

(4)
```
  1 2 5
+   4 3
```

(7)
```
  1 1 7
+     6
```

(10)
```
  1 2 6
+   5 6
```

(2)
```
  1 0 7
+   2 0
```

(5)
```
  1 3 5
+   4 5
```

(8)
```
  1 2 7
+     9
```

(3)
```
  1 3 0
+   3 6
```

(6)
```
  1 2 0
+     6
```

(9)
```
  1 3 7
+     4
```

2 Add.

5 points per question

(1)
```
  1 0 0
+ 1 0 0
```

(4)
```
  2 3 6
+ 1 2 3
```

(7)
```
  3 2 4
+ 4 2 7
```

(10)
```
  4 0 8
+ 1 3 5
```

(2)
```
  2 0 0
+ 2 0 0
```

(5)
```
  2 3 6
+ 1 5 2
```

(8)
```
  3 2 4
+ 4 2 9
```

(3)
```
  3 0 0
+ 3 0 0
```

(6)
```
  2 3 6
+ 1 2 9
```

(9)
```
  1 3 5
+ 3 1 7
```

Keep up the good work!

Reading
DAY
1

Vocabulary
Prepositions

Level ☆

Score

Date	Name
7 / 1 / 20	maddie

100/100

① Read each sentence aloud. Then trace the preposition.

5 points per question

(1) We walk _to_ the park.

(2) The monkey swings _on_ the branch.

(3) She met me _at_ the pool.

(4) The cat hid _under_ the couch.

(5) The dog jumps _in_ the lake.

(6) He ran _from_ the bookstore.

(7) We wait _for_ the bus.

(8) They eat the box _of_ candy.

(9) The birds perch _near_ the fountain.

(10) We sit _beside_ our camp counselor.

② Circle the preposition in each sentence below.

5 points per question

(1) I jogged (beside) my brother.

(2) We danced (to) the music.

(3) She jumped (on) the diving board.

(4) The man asked (for) three hot dogs.

(5) The turtle swam (under) the bridge.

(6) (In) the rollerskating rink, the skater spun around.

(7) He sunk a basket (from) mid-court.

(8) (At) the field, the team captains chose teammates.

(9) He wrote a book (of) short stories.

(10) (Near) the finish line, people cheered.

> **Don't forget!**
> A **preposition** is a word such as "with" or "on" that shows the relation of a noun or pronoun to other items in the sentence.

Date
7 / 1 / 20

Name
Maddie

100/100

① Add.

5 points per question

(1)
```
   1 6 2
 + 1 5 4
   3 1 6
```

(4)
```
   5 8 0
 + 1 5 0
   7 3 0
```

(7)
```
   2 7 5
 + 3 0 5
   5 8 0
```

(10)
```
   2 7 5
 +   3 6
   3 1 0    3 1 1
```

(2)
```
   2 5 1
 + 1 9 0
   4 4 1
```

(5)
```
   3 6 0
 + 3 8 2
   7 4 2    7 4 0
```

(8)
```
   2 7 5
 +   1 6
   2 9 1
```

(3)
```
   1 3 6
 + 2 7 2
   4 0 8
```

(6)
```
   2 5 0
 + 3 7 1
   6 2 1
```

(9)
```
   2 7 5
 +   2 6
   3 0 1
```

② Add.

5 points per question

(1)
```
   2 2 9
 +   7 6
   3 0 5
```

(4)
```
   2 9 3
 + 3 9 4
   6 8 7
```

(7)
```
   3 7 5
 + 4 8 7
   8 6 2
```

(10)
```
   1 4 1
 + 2 5 9
   4 0 0
```

(2)
```
   3 1 7
 +   8 5
   4 0 2
```

(5)
```
   2 9 8
 + 3 9 4
   6 9 2
```

(8)
```
   3 6 7
 + 2 6 2
   6 2 9
```

(3)
```
   2 6 4
 +   6 7
   3 3 1
```

(6)
```
   4 7 5
 + 3 9 6
   8 7 1
```

(9)
```
   1 7 8
 + 2 3 9
   4 1 7
```

Reading
DAY
2

Vocabulary
Dictionary skills

Level

Score

Date
7 / 2 / 2ₐ

Name
maddie

100   /100

① Trace the words in the "How to Read a Dictionary" passage below and read the definitions.   40 points for complet

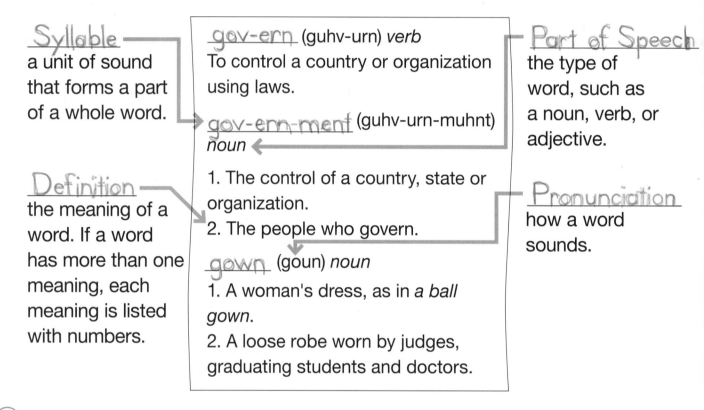

Syllable — a unit of sound that forms a part of a whole word.

gov-ern (guhv-urn) *verb*
To control a country or organization using laws.

gov-ern-ment (guhv-urn-muhnt) *noun*

Part of Speech — the type of word, such as a noun, verb, or adjective.

Definition — the meaning of a word. If a word has more than one meaning, each meaning is listed with numbers.

1. The control of a country, state or organization.
2. The people who govern.

gown (goun) *noun*
1. A woman's dress, as in *a ball gown*.
2. A loose robe worn by judges, graduating students and doctors.

Pronunciation — how a word sounds.

② Choose words from the box below to complete the descriptions.
Hint: You can use a word more than once.   12 points per question

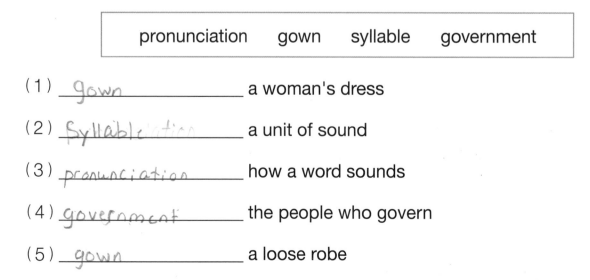

| pronunciation   gown   syllable   government |

(1) __gown_____ a woman's dress

(2) __Syllable ation_____ a unit of sound

(3) __pronunciation_____ how a word sounds

(4) __government_____ the people who govern

(5) __gown_____ a loose robe

4

Level ☆

Score

Date 7/2/20

Name maddie

100 /100

1 Subtract.

5 points per question

(1)
```
  o 12
  1 2 3
-   6 2
    6 1
```

(4)
```
  1 4 0
-   1 0
  1 3 0
```

(7)
```
  1 5 4
-   3 2
  1 2 2
```

(10)
```
    6 14
  1 7 4
-   1 8
  1 5 6
```

(2)
```
      14
  o 15 16
  1 5 0
-   7 2
    7 8
```

(5)
```
  1 6 0
-   4 0
  1 2 0
```

(8)
```
  1 5 6
-   5 6
  1 0 0
```

(3)
```
  o 13
  1 3 8
-   5 3
    8 5
```

(6)
```
  1 8 0
-   3 0
  1 5 0
```

(9)
```
    4 16
  1 5 6
-   4 8
  1 0 8
```

2 Subtract.

5 points per question

(1)
```
    3 15
  1 4 5
-   3 8
  1 0 7
```

(4)
```
    3 12
  3 4 2
-   3 8
  3 0 4
```

(7)
```
  6 0 0
- 2 0 0
  4 0 0
```

(10)
```
  7 4 4
- 4 1 0
  3 3 4
```

(2)
```
    3 15
  2 4 5
-   3 8
  2 0 7
```

(5)
```
  3 16
  4 6 7
-   8 7
  3 8 0
```

(8)
```
  7 4 0
- 3 0 0
  4 4 0
```

(3)
```
  o 14
  1 4 6
-   5 2
    9 4
```

(6)
```
    5 17
  4 6 7
-   5 9
  4 0 8
```

(9)
```
  7 4 0
- 3 2 0
  4 2 0
```

You're doing great!

Reading
DAY
3

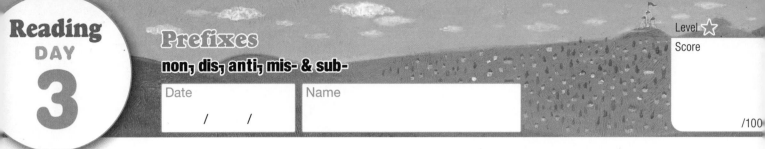

Prefixes

non-, dis-, anti-, mis- & sub-

Date
/ /

Name

Level ☆
Score

/100

① Trace the prefixes below. Then write the new words according to the   10 points per question
example.

(1) non + slip = nonslip

(2) dis + like = dislike

(3) anti + toxic = antitoxic

(4) sub + title = Subtitle

(5) mis + step = misstep

② Make words to match the definitions below by connecting two puzzle   10 points per question
pieces.

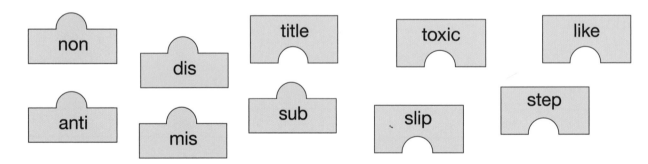

non

dis

title

toxic

like

anti

mis

sub

slip

step

(1) nonslip   created to reduce or prevent slipping

(2) misstep   a wrong step

(3) antitoxic   made to prevent, reduce or stop toxins

(4) Subtitle   a secondary title that explains more about the story

(5) dislike   not like something; feel disgust for something

Three-Digit Subtraction

Level ☆

Score

Math
DAY
4

/100

Date / /

Name

1 Subtract.

5 points per question

(1)
```
   4 3 5
 - 3 1 3
 ───────
   1 2 2
```

(4)
```
   5 5 6
 - 3 5 2
 ───────
   2 0 4
```

(7)
```
   ³ ¹⁴
   4 4 4
 - 1 6 2
 ───────
   2 8 2
```

(10)
```
   ⁵ ¹³
   6 3 6
 - 5 7 2
 ───────
     6 4
```

(2)
```
   4 3 5
 - 2 2 5
 ───────
   2 1 0
```

(5)
```
   5 5 6
 - 4 4 3
 ───────
   1 1 3
```

(8)
```
     ³ ¹⁴
   4 4 4
 - 1 2 6
 ───────
   3 1 8
```

(3)
```
   4 3 5
 - 4 0 0
 ───────
     3 5
```

(6)
```
     ⁴ ¹⁶
   5 5 6
 - 3 2 8
 ───────
   2 2 8
```

(9)
```
   4 4 4
 - 1 4 1
 ───────
   3 0 3
```

2 Subtract.

5 points per question

(1)
```
   ² ¹²
   1 3 2
 -   1 5
 ───────
   1 1 7
```

(4)
```
   ¹ ¹¹ ¹⁵
   2 2 5
 -   4 8
 ───────
   1 7 7
```

(7)
```
   1 0 0
 -     2
 ───────
     9 8 ¹⁰
```

(10)
```
   1 0 1
 -     2
 ───────
     9 9
```

(2)
```
   ² ¹²
   1 3 2
 -   2 7
 ───────
   1 0 5
```

(5)
```
     ¹
   2 2 5
 -   3 8
 ───────
   1 8 7
```

(8)
```
   2 0 0
 -     2
 ───────
   1 9 8
```

(3)
```
   ⁰ ² ¹²
   1 3 2
 -   3 8
 ───────
     9 4
```

(6)
```
   2 2 5
 - 1 5 8
 ───────
     6 7
```

(9)
```
   2 1 0
 -     5
 ───────
   2 0 5
```

Reading
DAY
4

Suffixes
-ly

Date
/ /

Name

Level ☆
Score

/100

① An adjective describes a noun. An adverb describes a verb. Complete 30 points for completion
the table below according to the example.

adjective	adverb
bright	brightly
recent	recently
loud	loudly
clever	cleverly
wise	wisely
brave	bravely
gentle	gently

Don't forget!
Adverbs usually have "ly" as a suffix.

② Complete the sentences with a pair of words from the box below. 10 points per question

> gentle/gently   recent/recently   wise/wisely   clever/cleverly
> loud/loudly   bright/brightly   brave/bravely

(1) My brother is __wise__. He __wisely__ eats many vegetables.

(2) The band is __loud__. The guitarist plays his solo __loudly__.

(3) In __recent__ days, the baby grew new teeth. She __recently__ started eating solid food.

(4) The mother __gently__ licked her kitten. The kitten purred because of the __gentle__ strokes.

(5) The moon was __bright__. It __brightly__ lit the field.

(6) What a __clever__ student! He __cleverly__ solved the math problem.

(7) The __brave__ firefighter fought the flames. She __bravely__ saved the family's home.

Multiplying 2 & 3

Level ☆
Score

Math
DAY
5

/100

Date / /

Name

1 Fill in the missing multiples of 2 in the boxes below.

14 points for completion

2 – 4 – 6 – 8 – 10 – 12 – 14 – 16 – 18

2 Say the number sentence aloud as you multiply by 2.

4 points per question

(1) 2 × 1 = 2
Two times one equals two

(6) 2 × 6 = 12
Two times six equals twelve

(2) 2 × 2 = 4
Two times two equals four

(7) 2 × 7 = 14
Two times seven equals fourteen

(3) 2 × 3 = 6
Two times three equals six

(8) 2 × 8 = 16
Two times eight equals sixteen

(4) 2 × 4 = 8
Two times four equals eight

(9) 2 × 9 = 18
Two times nine equals eighteen

(5) 2 × 5 = 10
Two times five equals ten

3 Fill in the missing multiples of 3 in the boxes below.

14 points for completion

3 – 6 – 9 – 12 – 15 – 18 – 21 – 24 – 27

4 Say the number sentence aloud as you multiply by 3.

4 points per question

(1) 3 × 1 = 3
Three times one equals three

(6) 3 × 6 = 18
Three times six equals eighteen

(2) 3 × 2 = 6
Three times two equals six

(7) 3 × 7 = 21
Three times seven equals twenty-one

(3) 3 × 3 = 9
Three times three equals nine

(8) 3 × 8 = 24
Three times eight equals twenty-four

(4) 3 × 4 = 12
Three times four equals twelve

(9) 3 × 9 = 27
Three times nine equals twenty-seven

(5) 3 × 5 = 15
Three times five equals fifteen

Reading
DAY
5

Compare

Date　　/　　/

Name

① Read the passage. Then match the words in bold with the words with similar meaning below. 10 points per question

It was a **dismal** day and Gary was bored. He thought it would be **amusing** to scare his little sister. He hid in her closet, and when she came in, he jumped out and yelled "Boo!" She screamed and fell back and scraped her arm. Gary didn't mean to **harm** her, but she began to cry. He got a bandage to help **mend** the scrape. He apologized and she **responded**, "I forgive you but watch out—I'm going to get even!" Gary now lived in fear! He surely wasn't bored anymore.

(1) reply 　(responded)　　　(2) fix 　(mend)

(3) funny 　(amusing)　　　(4) gloomy 　(dismal)

(5) hurt 　(harm)

② Read the sentence then put a (✓) next to the phrase that is similar. 10 points per question

(1) interested in books

(✓) likes to read 　(✗) thinks about movies

(2) shares her treat

(✓) offers her popcorn 　(✗) eats her own lunch

(3) breaks it in two

(✓) snaps it in half 　(✗) smashes it to pieces

(4) slips on the grass

(✗) sits on the grass 　(✓) slides on the grass

(5) dark hair and eyes

(✓) black hair and brown eyes 　(✗) black hair and blue eyes

Level ☆
Score

/100

Date / /

Name

1 Fill in the missing multiples of 4 in the boxes below.

14 points for completion

$4 - 8 - 12 - 16 - 20 - 24 - 28 - 32 - 36$

2 Say the number sentence aloud as you multiply by 4.

4 points per question

(1) $4 \times 1 = \boxed{4}$
Four times one equals four

(6) $4 \times 6 = \boxed{24}$
Four times six equals twenty-four

(2) $4 \times 2 = \boxed{8}$
Four times two equals eight

(7) $4 \times 7 = \boxed{28}$
Four times seven equals twenty-eight

(3) $4 \times 3 = \boxed{12}$
Four times three equals twelve

(8) $4 \times 8 = \boxed{32}$
Four times eight equals thirty-two

(4) $4 \times 4 = \boxed{16}$
Four times four equals sixteen

(9) $4 \times 9 = \boxed{36}$
Four times nine equals thirty-six

(5) $4 \times 5 = \boxed{20}$
Four times five equals twenty

3 Fill in the missing multiples of 5 in the boxes below.

14 points for completion

$5 - 10 - 15 - 20 - 25 - 30 - 35 - 40 - 45$

4 Say the number sentence aloud as you multiply by 5.

4 points per question

(1) $5 \times 1 = \boxed{5}$
Five times one equals five

(6) $5 \times 6 = \boxed{30}$
Five times six equals thirty

(2) $5 \times 2 = \boxed{10}$
Five times two equals ten

(7) $5 \times 7 = \boxed{35}$
Five times seven equals thirty-five

(3) $5 \times 3 = \boxed{15}$
Five times three equals fifteen

(8) $5 \times 8 = \boxed{40}$
Five times eight equals forty

(4) $5 \times 4 = \boxed{20}$
Five times four equals twenty

(9) $5 \times 9 = \boxed{45}$
Five times nine equals forty-five

(5) $5 \times 5 = \boxed{25}$
Five times five equals twenty-five

Reading
DAY
6

Contrast

Level ☆

Score

Date / /

Name

/100

① Read the passage. Then match the words in bold with the words with opposite meanings below. 10 points per question

My family was planning a tubing trip down the Delaware River. My dad wanted to **purchase** an **enormous** raft for himself so he could stretch out. He searched and searched for the **perfect** raft. He found a huge raft that came with a **powerful** pump. Even so, when we went to the river, the raft took forever to blow up. When we finally got into the water, our tubes floated quickly, but my dad's heavy raft got stuck on every rock! As we **vanished** down the river, we yelled, "You forgot to buy paddles!"

(1) weak   (powerful)

(2) small   (enourmas)

(3) appeared   (vanished)

(4) sell   (purchase)

(5) faulty   (perfect)

② Read the sentence then put a check (✓) next to the contrasting phrase. 10 points per question

(1) likes to sing

(✗) enjoys being in a chorus   (✓) loves to play piano

(2) bakes some bread

(✓) eats some bread   (✗) puts bread in oven

(3) glues the pieces together

(✓) breaks it apart   (✗) pastes the parts

(4) bikes fast

(✗) speeds away   (✓) stands to watch

(5) bushy tail wags

(✗) hairy tail thumps   (✓) dog sleeps soundly

Multiplying 6 & 7

Score

/100

Date   /   /

Name

1 Fill in the missing multiples of 6 in the boxes below.

14 points for completion

| 6 – 12 – 18 – 24 – 30 – 36 – 42 – 48 – 54 |

2 Say the number sentence aloud as you multiply by 6.

4 points per question

(1) 6 × 1 = 6
Six times one equals six

(2) 6 × 2 = 12
Six times two equals twelve

(3) 6 × 3 = 18
Six times three equals eighteen

(4) 6 × 4 = 24
Six times four equals twenty-four

(5) 6 × 5 = 30
Six times five equals thirty

(6) 6 × 6 = 36
Six times six equals thirty-six

(7) 6 × 7 = 42
Six times seven equals forty-two

(8) 6 × 8 = 48
Six times eight equals forty-eight

(9) 6 × 9 = 54
Six times nine equals fifty-four

3 Fill in the missing multiples of 7 in the boxes below.

14 points for completion

| 7 – 14 – 21 – 28 – 35 – 42 – 49 – 56 – 63 |

4 Say the number sentence aloud as you multiply by 7.

4 points per question

(1) 7 × 1 = 7
Seven times one equals seven

(2) 7 × 2 = 14
Seven times two equals fourteen

(3) 7 × 3 = 21
Seven times three equals twenty-one

(4) 7 × 4 = 28
Seven times four equals twenty-eight

(5) 7 × 5 = 35
Seven times five equals thirty-five

(6) 7 × 6 = 42
Seven times six equals forty-two

(7) 7 × 7 = 49
Seven times seven equals forty-nine

(8) 7 × 8 = 56
Seven times eight equals fifty-six

(9) 7 × 9 = 63
Seven times nine equals sixty-three

Reading
DAY
7

Defining Words by Context
Belling the Cat 1

Level ☆
Score

/100

Date
/   /

Name

1 Read the passage. Then choose the words in bold from the passages **10** points per question
to complete the definitions below.

Long ago, some house mice met to **consider** what **measures** they could take to **outsmart** their **enemy**, the cat. Some said this, and some said that. At last a young mouse got up and said he had a **proposal** to make, that he thought would meet the **challenge**.

"You will all agree," said he, "that our **chief** danger is the **sly** way in which the enemy attacks us. Now, if we could receive some **signal** that she is coming, we could easily **escape** from her."

(1) __proposal__   a suggestion

(2) __sly__   clever, tricky, or sneaky

(3) __signal__   a sound or motion made to give a warning or an order

(4) __challenge__   a difficult task or a contest

(5) __cheif__   most important

(6) __measures__   plans or courses of action

(7) __consider__   to think carefully about something

(8) __outsmart__   defeat by being clever

(9) __enemy__   opponent or rival

(10) __escape__   break free

Multiplying 8 & 9

Level ☆

Score

Math
DAY
8

/100

Date / /

Name

1 Fill in the missing multiples of 8 in the boxes below.

14 points for completion

$8 - 16 - 24 - 32 - 40 - 48 - 56 - 64 - 72$

2 Say the number sentence aloud as you multiply by 8.

4 points per question

(1) $8 \times 1 = 8$
Eight times one equals eight

(2) $8 \times 2 = 16$
Eight times two equals sixteen

(3) $8 \times 3 = 24$
Eight times three equals twenty-four

(4) $8 \times 4 = 32$
Eight times four equals thirty-two

(5) $8 \times 5 = 40$
Eight times five equals forty

(6) $8 \times 6 = 48$
Eight times six equals forty-eight

(7) $8 \times 7 = 56$
Eight times seven equals fifty-six

(8) $8 \times 8 = 64$
Eight times eight equals sixty-four

(9) $8 \times 9 = 72$
Eight times nine equals seventy-two

3 Fill in the missing multiples of 9 in the boxes below.

14 points for completion

$9 - 18 - 27 - 36 - 45 - 54 - 63 - 72 - 81$

4 Say the number sentence aloud as you multiply by 9.

4 points per question

(1) $9 \times 1 = 9$
Nine times one equals nine

(2) $9 \times 2 = 18$
Nine times two equals eighteen

(3) $9 \times 3 = 27$
Nine times three equals twenty-seven

(4) $9 \times 4 = 36$
Nine times four equals thirty-six

(5) $9 \times 5 = 45$
Nine times five equals forty-five

(6) $9 \times 6 = 54$
Nine times six equals fifty-four

(7) $9 \times 7 = 63$
Nine times seven equals sixty-three

(8) $9 \times 8 = 72$
Nine times eight equals seventy-two

(9) $9 \times 9 = 81$
Nine times nine equals eighty-one

Reading
DAY
8

Defining Words by Context
Belling the Cat 2

Date   /   /

Name

Level ☆
Score

/100

① Read the passage. Then choose words from the passage to complete the definitions below.

10 points per question

> The young mouse went on to **explain** how they would know if the cat was nearby.
> "A small bell should be **acquired**, and **attached** with a ribbon around the neck of the cat. Then we will always know when she is around, and can easily **retreat** while she is in the **neighborhood**."
> All the mice **applauded** for the smart **scheme**, except for the old mouse. He got up and said, "That is all very well, but who will bell the cat?"
> The mice looked at one another and **grumbled** their excuses: "I have a bad back!" "I'm afraid of bells!" "I'm busy later."
> Even the young mouse said, "I came up with the idea! I shouldn't have to bell the cat!" Then all the mice fell silent and the old mouse **concluded**,
> "It is easy to think of **impossible** solutions."

(1) _neighboorhood_   an area around a place, person or object

(2) _Concluded_   judged or arrived at a judgement

(3) _grumbled_   complained, whined or protested

(4) _attached_   joined or connected to something or someone

(5) _retreat_   to back down or withdraw from an enemy

(6) _acquired_   bought or gotten

(7) _impossible_   not able to occur, exist or be done

(8) _Scheme_   a plan

(9) _applauded_   clapped

(10) _explain_   make an idea clear by describing it in detail

Multiplying 1 & 10

Level ☆
Score
/100

Date   /   /

Name

---

1 Fill in the missing multiples of 1 in the boxes below.

14 points for completion

| 1 | – | 2 | – | 3 | – | 4 | – | 5 | – | 6 | – | 7 | – | 8 | – | 9 |

2 Say the number sentence aloud as you multiply by 1.

4 points per question

(1) 1 × 1 = 1
One times one equals one

(2) 1 × 2 = 2
One times two equals two

(3) 1 × 3 = 3
One times three equals three

(4) 1 × 4 = 4
One times four equals four

(5) 1 × 5 = 5
One times five equals five

(6) 1 × 6 = 6
One times six equals six

(7) 1 × 7 = 7
One times seven equals seven

(8) 1 × 8 = 8
One times eight equals eight

(9) 1 × 9 = 9
One times nine equals nine

---

3 Fill in the missing multiples of 10 in the boxes below.

14 points for completion

| 10 | – | 20 | – | 30 | – | 40 | – | 50 | – | 60 | – | 70 | – | 80 | – | 90 |

4 Say the number sentence aloud as you multiply by 10.

4 points per question

(1) 10 × 1 = 10
Ten times one equals ten

(2) 10 × 2 = 20
Ten times two equals twenty

(3) 10 × 3 = 30
Ten times three equals thirty

(4) 10 × 4 = 40
Ten times four equals forty

(5) 10 × 5 = 50
Ten times five equals fifty

(6) 10 × 6 = 60
Ten times six equals sixty

(7) 10 × 7 = 70
Ten times seven equals seventy

(8) 10 × 8 = 80
Ten times eight equals eighty

(9) 10 × 9 = 90
Ten times nine equals ninety

Reading

DAY

9

Defining Words by Context

The Sweet Soup 1

Date / /

Name

Level ⭐⭐

Score

/100

1 Read the passage and the vocabulary words defined below. Complete the passage using the vocabulary words.

10 points per question

Once upon a time there was a poor but very good little girl who lived alone with her mother. They had nothing in the house to eat.

So the child went to (1) _forage_ in the forest, and there she met an (2) _elderly_ woman. The old woman saw she was (3) _frail_ because the girl looked skinny and (4) _disstressed_. The old woman (5) _presented_ her with a pot which had (6) _magical_ powers. If one said to it, "Boil, little pot!" it would cook sweet soup; and when one (7) _commanded_, "Stop, little pot!" it would (8) _immediately_ stop boiling. The little girl took the pot home to her mother, and now their (9) _poverty_ was at an end, for they could have sweet broth (10) _whenever_ they wanted.

distress: in extreme pain or sadness

forage: search widely for food, drink or needed items

frail: weak and delicate

immediately: at once; instantly

magical: using magic

poverty: the state of being very poor

presented: gave or offered

whenever: at any time

elderly: an old or aging person

commanded: gave an order

Multiplication

Level ★★
Score

Math
DAY
10

/100

1 Multiply.

(1) $2 \times 2 = 4$

(2) $2 \times 4 = 8$

(3) $2 \times 6 = 12$

(4) $2 \times 8 = 16$

(5) $2 \times 9 = 18$

(6) $3 \times 5 = 15$

(7) $3 \times 6 = 18$

(8) $3 \times 7 = 21$

(9) $3 \times 2 = 6$

(10) $3 \times 8 = 24$

(11) $4 \times 3 = 12$

(12) $4 \times 5 = 20$

(13) $4 \times 7 = 28$

(14) $4 \times 9 = 36$

(15) $4 \times 2 = 8$

(16) $5 \times 2 = 10$

(17) $5 \times 4 = 20$

(18) $5 \times 5 = 25$

(19) $5 \times 7 = 35$

(20) $5 \times 9 = 45$

(21) $6 \times 4 = 24$

(22) $6 \times 5 = 30$

(23) $6 \times 6 = 36$

(24) $6 \times 8 = 48$

(25) $6 \times 3 = 18$

(26) $7 \times 3 = 21$

(27) $7 \times 5 = 35$

(28) $7 \times 6 = 42$

(29) $7 \times 8 = 55$

(30) $7 \times 9 = 63$

(31) $8 \times 3 = 24$

(32) $8 \times 5 = 40$

(33) $8 \times 7 = 55$

(34) $8 \times 9 = 72$

(35) $8 \times 3 = 24$

(36) $9 \times 5 = 45$

(37) $9 \times 6 = 54$

(38) $9 \times 7 = 63$

(39) $9 \times 4 = 36$

(40) $9 \times 9 = 81$

(41) $1 \times 2 = 2$

(42) $1 \times 4 = 4$

(43) $1 \times 6 = 6$

(44) $1 \times 8 = 8$

(45) $1 \times 9 = 9$

(46) $10 \times 3 = 30$

(47) $10 \times 5 = 50$

(48) $10 \times 7 = 70$

(49) $10 \times 8 = 80$

(50) $10 \times 9 = 90$

Defining Words by Context
The Sweet Soup 2

Date   /   /

Name

1 Read the passage and the vocabulary words defined below. Complete the passage using the vocabulary words.

10 points per question

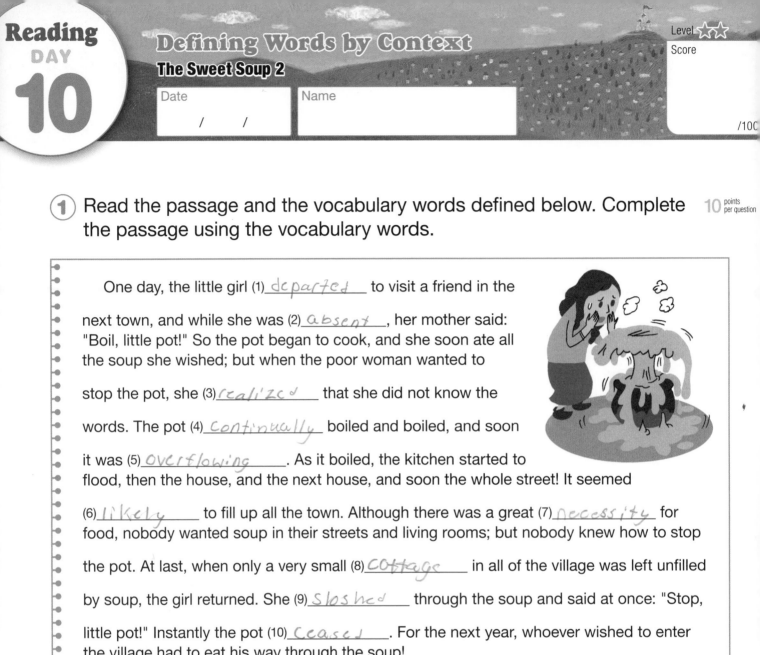

One day, the little girl (1) _departed_ to visit a friend in the next town, and while she was (2) _absent_, her mother said: "Boil, little pot!" So the pot began to cook, and she soon ate all the soup she wished; but when the poor woman wanted to stop the pot, she (3) _realized_ that she did not know the words. The pot (4) _continually_ boiled and boiled, and soon it was (5) _overflowing_. As it boiled, the kitchen started to flood, then the house, and the next house, and soon the whole street! It seemed (6) _likely_ to fill up all the town. Although there was a great (7) _necessity_ for food, nobody wanted soup in their streets and living rooms; but nobody knew how to stop the pot. At last, when only a very small (8) _cottage_ in all of the village was left unfilled by soup, the girl returned. She (9) _sloshed_ through the soup and said at once: "Stop, little pot!" Instantly the pot (10) _ceased_. For the next year, whoever wished to enter the village had to eat his way through the soup!

absent: away or not present

ceased: came to an end

cottage: a small simple house

likely: probable or expected

necessity: a required or needed thing

overflowing: flooding or flowing over; so full that the contents spills over the rim

realized: became aware; noticed

sloshed: moved through liquid with a splashing sound

departed: went away

continually: constantly; again and again

Level ★★

Score

/100

Math
DAY
11

1 Multiply.

2 points per question

(1) $3 \times 4 = 12$

(2) $2 \times 9 = 18$

(3) $2 \times 1 = 2$

(4) $3 \times 7 = 21$

(5) $2 \times 8 = 16$

(6) $3 \times 3 = 9$

(7) $3 \times 8 = 24$

(8) $2 \times 5 = 10$

(9) $3 \times 2 = 6$

(10) $3 \times 6 = 18$

(11) $4 \times 6 = 24$

(12) $5 \times 1 = 5$

(13) $5 \times 3 = 15$

(14) $4 \times 9 = 36$

(15) $5 \times 2 = 10$

(16) $4 \times 5 = 20$

(17) $4 \times 6 = 24$

(18) $5 \times 7 = 35$

(19) $4 \times 4 = 16$

(20) $4 \times 8 = 32$

(21) $7 \times 4 = 28$

(22) $6 \times 10 = 60$

(23) $6 \times 1 = 6$

(24) $7 \times 7 = 46$

(25) $6 \times 2 = 12$

(26) $7 \times 3 = 21$

(27) $7 \times 8 = 56$

(28) $6 \times 5 = 30$

(29) $7 \times 2 = 14$

(30) $7 \times 6 = 42$

(31) $9 \times 6 = 54$

(32) $8 \times 9 = 72$

(33) $8 \times 1 = 8$

(34) $9 \times 7 = 63$

(35) $8 \times 3 = 24$

(36) $9 \times 3 = 27$

(37) $9 \times 8 = 72$

(38) $8 \times 5 = 40$

(39) $9 \times 2 = 18$

(40) $9 \times 6 = 54$

(41) $10 \times 6 = 60$

(42) $1 \times 1 = 1$

(43) $1 \times 3 = 3$

(44) $10 \times 9 = 90$

(45) $1 \times 2 = 2$

(46) $10 \times 5 = 50$

(47) $10 \times 4 = 40$

(48) $1 \times 7 = 7$

(49) $10 \times 2 = 20$

(50) $10 \times 8 = 80$

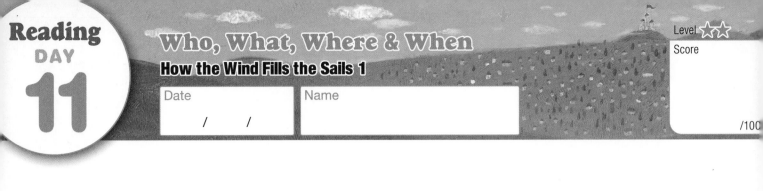

Reading
DAY
11

Who, What, Where & When

How the Wind Fills the Sails 1

Level ⭐⭐
Score

/100

Date / /

Name

① Read the passage from "How the Wind Fills the Sails" by Dora Bernside. 100 points for complete
Use words from the passage to answer the questions below.

> "What makes the vessel move on the river?" asked little Anna one day of her brother Harry.
>
> "Why," said Harry, "it's the wind, of course, that fills the sails, and that pushes the vessel on. Come out on the lawn, and I will show you how it is done."
>
> So Anna, Harry, and Bravo, all ran out on the lawn. Bravo was a dog, but he was always curious to see what was going on.
>
> When they were on the lawn, Harry took out his handkerchief and told Anna to hold it by two of the corners while he held the other two corners.
>
> As soon as they had done this, the wind made it swell out, and look just like a sail.
>
> "Now you see how the wind fills the sails," said Harry.

(1) Who wants to know what makes vessels move on the river?

___Anna___ **wants to know what makes vessels move on the river.**

(2) What pushes the vessel on the river?

The ___wind___ pushes the vessel on the river.

(3) Where does Harry suggest they go to test his idea?

Harry suggests they go out on the ___lawn___.

(4) When does Harry take out his handkerchief?

Harry takes out his handkerchief when they are on the ___lawn___.

(5) Who is always curious to see what is going on?

___Bravo___ **is always curious to see what is going on.**

(6) What makes the handkerchief swell out?

The ___wind___ makes the handkerchief swell out.

(7) Where do Anna and Harry hold the handkerchief?

Anna and Harry hold the handkerchief at the ___corners___

(8) When the wind makes the handkerchief swell out, what does it look like?

The handkerchief looks like a ___sail___.

Level ★★

Score

/100

Date   /   /

Name

1 Multiply.

2 points per question

(1) $4 \times 2 = 8$

(2) $6 \times 5 = 30$

(3) $8 \times 1 = 8$

(4) $2 \times 6 = 12$

(5) $3 \times 4 = 12$

(6) $7 \times 7 = 49$

(7) $5 \times 1 = 5$

(8) $9 \times 3 = 27$

(9) $3 \times 8 = 24$

(10) $5 \times 3 = 15$

(11) $2 \times 1 = 2$

(12) $1 \times 6 = 6$

(13) $7 \times 8 = 56$

(14) $9 \times 2 = 18$

(15) $2 \times 4 = 8$

(16) $8 \times 8 = 64$

(17) $6 \times 9 = 54$

(18) $4 \times 7 = 28$

(19) $8 \times 5 = 40$

(20) $3 \times 3 = 9$

(21) $7 \times 5 = 35$

(22) $6 \times 8 = 48$

(23) $4 \times 4 = 16$

(24) $10 \times 8 = 80$

(25) $2 \times 5 = 10$

(26) $8 \times 6 = 48$

(27) $3 \times 7 = 21$

(28) $4 \times 9 = 36$

(29) $9 \times 8 = 72$

(30) $6 \times 2 = 12$

(31) $7 \times 3 = 21$

(32) $5 \times 5 = 25$

(33) $2 \times 2 = 4$

(34) $8 \times 4 = 32$

(35) $9 \times 5 = 45$

(36) $6 \times 7 = 42$

(37) $2 \times 8 = 16$

(38) $3 \times 2 = 6$

(39) $4 \times 6 = 24$

(40) $7 \times 9 = 63$

(41) $9 \times 1 = 9$

(42) $2 \times 3 = 6$

(43) $8 \times 7 = 56$

(44) $5 \times 2 = 10$

(45) $1 \times 4 = 4$

(46) $10 \times 5 = 50$

(47) $4 \times 3 = 12$

(48) $3 \times 9 = 27$

(49) $7 \times 2 = 14$

(50) $9 \times 9 = 81$

Reading
DAY
12

Who, What, Where & When
How the Wind Fills the Sails 2

Level ⭐⭐
Score

Date
/   /

Name

/100

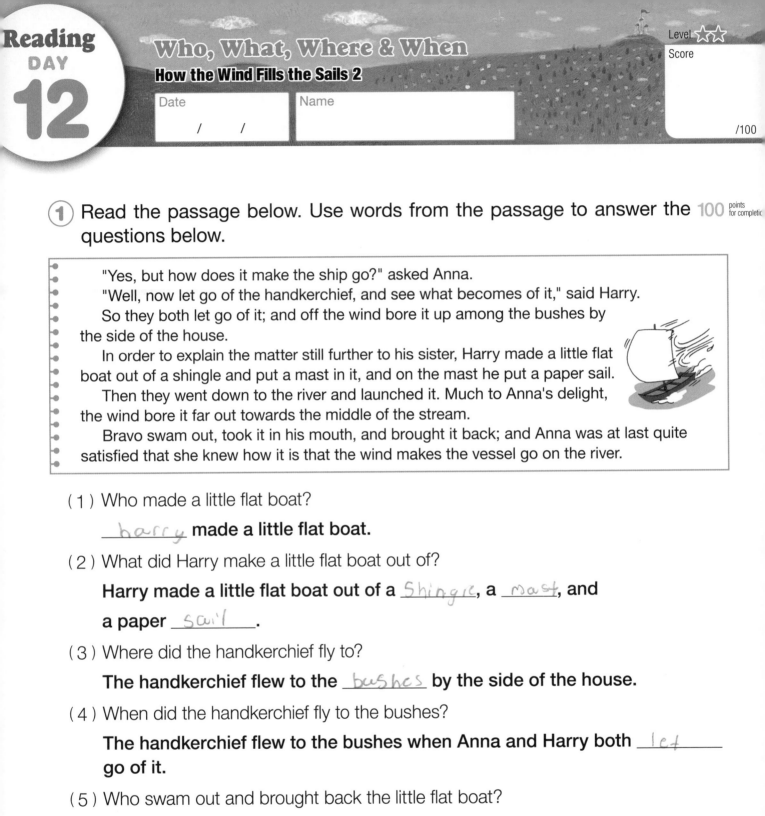

① Read the passage below. Use words from the passage to answer the 100 points for completion questions below.

"Yes, but how does it make the ship go?" asked Anna.

"Well, now let go of the handkerchief, and see what becomes of it," said Harry.

So they both let go of it; and off the wind bore it up among the bushes by the side of the house.

In order to explain the matter still further to his sister, Harry made a little flat boat out of a shingle and put a mast in it, and on the mast he put a paper sail.

Then they went down to the river and launched it. Much to Anna's delight, the wind bore it far out towards the middle of the stream.

Bravo swam out, took it in his mouth, and brought it back; and Anna was at last quite satisfied that she knew how it is that the wind makes the vessel go on the river.

(1) Who made a little flat boat?

____harry____ **made a little flat boat.**

(2) What did Harry make a little flat boat out of?

Harry made a little flat boat out of a _Shingle_**, a** _mast_**, and**

a paper _sail_ **.**

(3) Where did the handkerchief fly to?

The handkerchief flew to the _bushes_ **by the side of the house.**

(4) When did the handkerchief fly to the bushes?

The handkerchief flew to the bushes when Anna and Harry both _let_

go of it.

(5) Who swam out and brought back the little flat boat?

____Bravo____ **swam out and brought back the little flat boat.**

(6) Where do they go to launch the boat?

They go down to the _river_ **to launch the boat.**

(7) When was Anna convinced that she knew how vessels moved?

Anna was convinced when the wind bore the little ship out towards the

middle of the _Stream_**.**

Level ⭐⭐
Score

/100

Math
DAY
13

Date / /   Name

1 Write the missing number in each box.

3 points per question

(1) $2 \times \boxed{3} = 6$

(2) $2 \times \boxed{5} = 10$

(3) $2 \times \boxed{7} = 14$

(4) $3 \times \boxed{5} = 15$

(5) $4 \times \boxed{7} = 28$

(6) $5 \times \boxed{4} = 40$

(7) $6 \times \boxed{6} = 36$

(8) $7 \times \boxed{8} = 56$

(9) $8 \times \boxed{9} = 72$

(10) $9 \times \boxed{7} = 63$

(11) $10 \times \boxed{7} = 70$

(12) $\boxed{2} \times 2 = 4$

(13) $\boxed{3} \times 2 = 6$

(14) $\boxed{4} \times 3 = 12$

(15) $\boxed{4} \times 4 = 16$

(16) $\boxed{5} \times 5 = 25$

(17) $\boxed{3} \times 6 = 18$

(18) $\boxed{4} \times 7 = 28$

(19) $\boxed{8} \times 8 = 64$

(20) $\boxed{9} \times 9 = 81$

2 Look at the example. Then write the missing number in each box.

4 points per question

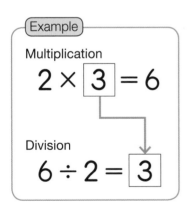

Example

Multiplication

$2 \times \boxed{3} = 6$

Division

$6 \div 2 = \boxed{3}$

(1) $2 \times \boxed{4} = 8$ $8 \div 2 = \boxed{4}$

(2) $2 \times \boxed{7} = 14$ $14 \div 2 = \boxed{7}$

(3) $3 \times \boxed{5} = 15$ $15 \div 3 = \boxed{5}$

(4) $4 \times \boxed{5} = 20$ $20 \div 4 = \boxed{5}$

(5) $5 \times \boxed{6} = 30$ $30 \div 5 = \boxed{6}$

(6) $6 \times \boxed{8} = 48$ $48 \div 6 = \boxed{8}$

(7) $7 \times \boxed{4} = 28$ $28 \div 7 = \boxed{4}$

(8) $8 \times \boxed{7} = 56$ $56 \div 8 = \boxed{7}$

(9) $9 \times \boxed{8} = 72$ $72 \div 9 = \boxed{8}$

(10) $10 \times \boxed{8} = 80$ $80 \div 10 = \boxed{8}$

Who, What, Where, When, Why & How

How Rabbit Fooled the Elephant and Whale 1

Date   /   /

Name

① Read the passage. Use words from the passage to answer the questions below. 100 points for comple

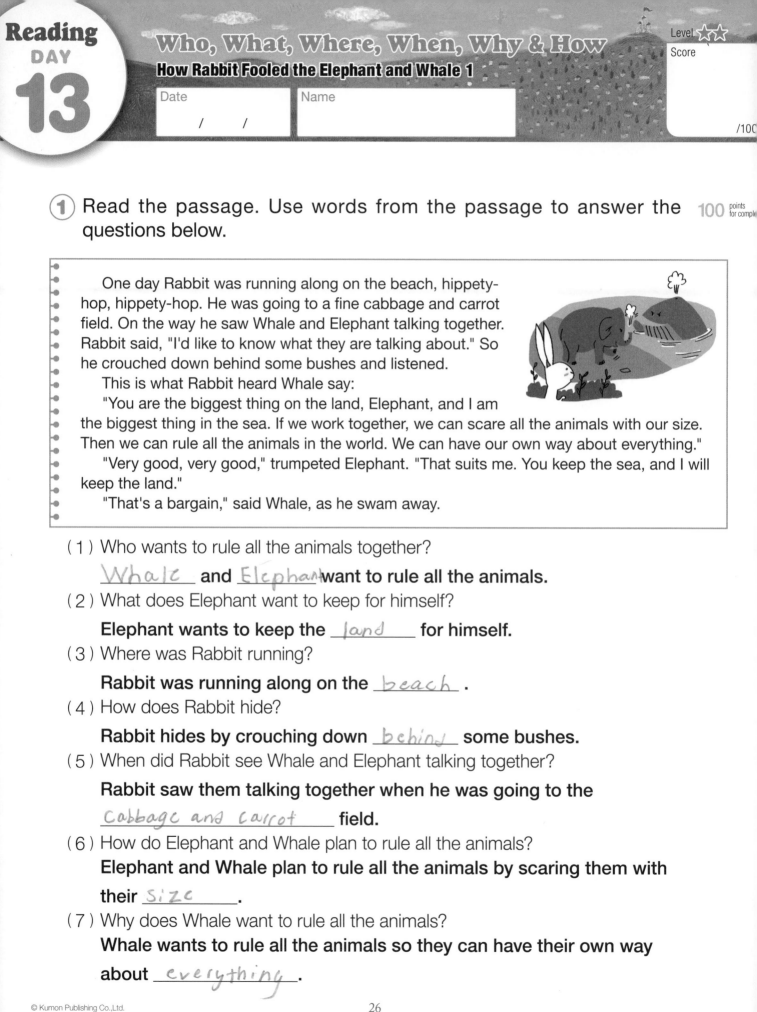

One day Rabbit was running along on the beach, hippety-hop, hippety-hop. He was going to a fine cabbage and carrot field. On the way he saw Whale and Elephant talking together. Rabbit said, "I'd like to know what they are talking about." So he crouched down behind some bushes and listened.

This is what Rabbit heard Whale say:

"You are the biggest thing on the land, Elephant, and I am the biggest thing in the sea. If we work together, we can scare all the animals with our size. Then we can rule all the animals in the world. We can have our own way about everything."

"Very good, very good," trumpeted Elephant. "That suits me. You keep the sea, and I will keep the land."

"That's a bargain," said Whale, as he swam away.

(1) Who wants to rule all the animals together?

Whale and _Elephant_ want to rule all the animals.

(2) What does Elephant want to keep for himself?

Elephant wants to keep the _land_ for himself.

(3) Where was Rabbit running?

Rabbit was running along on the _beach_ .

(4) How does Rabbit hide?

Rabbit hides by crouching down _behind_ some bushes.

(5) When did Rabbit see Whale and Elephant talking together?

Rabbit saw them talking together when he was going to the

Cabbage and Carrot field.

(6) How do Elephant and Whale plan to rule all the animals?

Elephant and Whale plan to rule all the animals by scaring them with

their _size_ .

(7) Why does Whale want to rule all the animals?

Whale wants to rule all the animals so they can have their own way

about _everything_ .

Division

Date   /   /

Name

1 Look at the example. Then write the missing number in each box.

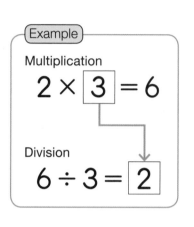

Example

Multiplication

$2 \times \boxed{3} = 6$

Division

$6 \div 3 = \boxed{2}$

(1) $\boxed{2} \times 2 = 4$ $4 \div 2 = \boxed{2}$

(2) $\boxed{6} \times 2 = 12$ $12 \div 2 = \boxed{6}$

(3) $\boxed{3} \times 3 = 9$ $9 \div 3 = \boxed{3}$

(4) $\boxed{4} \times 4 = 16$ $16 \div 4 = \boxed{4}$

(5) $\boxed{5} \times 5 = 25$ $25 \div 5 = \boxed{5}$

(6) $6 \times \boxed{7} = 42$ $42 \div 6 = \boxed{7}$

(7) $7 \times \boxed{5} = 35$ $35 \div 7 = \boxed{5}$

(8) $8 \times \boxed{6} = 48$ $48 \div 8 = \boxed{6}$

(9) $9 \times \boxed{7} = 63$ $63 \div 9 = \boxed{7}$

(10) $10 \times \boxed{6} = 60$ $60 \div 10 = \boxed{6}$

2 Divide.

(1) $8 \div 2 = 4$

(2) $10 \div 2 = 5$

(3) $18 \div 2 = 9$

(4) $6 \div 3 = 2$

(5) $21 \div 3 = 7$

(6) $8 \div 4 = 2$

(7) $24 \div 4 = 6$

(8) $10 \div 5 = 2$

(9) $40 \div 5 = 4$

(10) $12 \div 6 = 2$

(11) $42 \div 6 = 7$

(12) $49 \div 7 = 7$

(13) $56 \div 8 = 7$

(14) $54 \div 9 = 6$

(15) $70 \div 10 = 7$

Who, What, Where, When, Why & How

How Rabbit Fooled the Elephant and Whale 2

Date
/   /

Name

/100

① Read the passage. Use words from the passage to answer the   100 points for completi
questions below.

> Rabbit laughed to himself. "Elephant and Whale may be big, but they won't rule me," he said.
> Rabbit hopped off and soon came back with a very long and sturdy rope and his big drum. With one end of the rope, he walked up to Elephant who was munching on some tree leaves.
> "Oh, dear Mr. Elephant," he said, "you are so big and strong; will you have the kindness to do me a favor?"
> Elephant was flattered, and he trumpeted, "Certainly, certainly. What is it?"
> "My cow is stuck in the mud on the shore, and I can't pull her out," said Rabbit. "You are so strong, I am sure you can get her out."
> "Certainly, certainly," trumpeted Elephant.
> "Thank you," said Rabbit. "Take this rope in your trunk, and I will tie the other end to my cow. Then I will beat my drum to let you know when to pull. You must pull as hard as you can, for the cow is very heavy."
> "Huh!" trumpeted Elephant, "I'll pull her out by using my huge legs, or else I'll break the rope."
> Rabbit tied the rope to Elephant's trunk and hopped away.

(1) Who says that Elephant is big and strong?

_____Rabbit_____ **says that Elephant is big and strong.**

(2) What does Rabbit bring back with him?

Rabbit brings a very long and sturdy _rope_ **and his big** _drum_ **.**

(3) Where does Rabbit say he will tie the other end of the rope?

Rabbit says he will tie the other end to his _cow_ **.**

(4) When is Elephant supposed to pull?

Elephant is supposed to pull when he hears Rabbit _beat_ **the drum.**

(5) Why does Rabbit say he needs Elephant to pull his cow?

Rabbit says his _cow_ **is stuck in the** _mud_ **on the shore, and he can't pull her out.**

(6) How will Elephant pull out the cow?

Elephant will pull out the cow by using his huge _legs_ **.**

Division

Level ☆☆
Score

Math
DAY
15

/100

Date   /   /

Name

1 Divide.

4 points per question

(1) $8 \div 2 = 4$

(2) $12 \div 2 = 6$

(3) $10 \div 2 = 5$

(4) $18 \div 3 = 6$

(5) $24 \div 3 = 8$

(6) $6 \div 3 = 2$

(7) $20 \div 4 = 5$

(8) $32 \div 4 = 8$

(9) $12 \div 4 = 3$

(10) $20 \div 5 = 4$

(11) $35 \div 5 = 7$

(12) $40 \div 5 = 8$

(13) $30 \div 6 = 5$

(14) $6 \div 6 = 1$

(15) $42 \div 6 = 7$

(16) $28 \div 7 = 4$

(17) $14 \div 7 = 2$

(18) $42 \div 7 = 6$

(19) $32 \div 8 = 4$

(20) $56 \div 8 = 7$

(21) $64 \div 8 = 8$

(22) $27 \div 9 = 3$

(23) $54 \div 9 = 6$

(24) $63 \div 9 = 7$

(25) $50 \div 10 = 5$

Reading
DAY
15

Who, What, Where, When, Why & How

Level ★★
Score

How Rabbit Fooled the Elephant and Whale 3

Date   /   /

Name

/100

① Read the passage. Use words from the passage to answer the questions below.

100 points for comple

> Rabbit hopped till he came to the shore where Whale was. Making a bow, Rabbit said, "Mighty and wonderful Whale, will you do me a favor?"
>
> "What is it?" asked Whale.
>
> "My cow is stuck in the mud on the shore," said Rabbit, "and no one can pull her out. Of course you can do it. If you will be so kind as to help me, I will be very grateful."
>
> "Certainly," said Whale, "certainly."
>
> "Thank you," said Rabbit, "take hold of this rope in your teeth, and I will tie the other end to my cow. Then I will beat my big drum to let you know when to pull. You must pull as hard as you can, for my cow is very heavy."
>
> "Never fear," said Whale, "I could pull a dozen cows out of the mud by flapping my huge tail."
>
> "I am sure you could," said Rabbit politely. "Only be sure to begin gently. Then pull harder and harder till you get her out."

(1) Who does Rabbit call mighty and wonderful?

Rabbit calls _whale_ mighty and wonderful.

(2) What will Whale to hold in his teeth?

Whale will hold the _rope_ in his teeth.

(3) Where does Rabbit say he will tie the other end of the rope?

Rabbit says he will tie the other end to his _cow_ .

(4) When is Whale supposed to pull?

Whale is supposed to pull when he hears Rabbit _beat_ the drum.

(5) Why does Rabbit say he needs Whale to pull his cow?

Rabbit says his _cow_ is stuck in the _mud_ on the shore, and he can't pull her out.

(6) How will Whale pull out the cow?

Whale will pull out the cow by flapping his huge _tail_ .

30

Division

Level ☆☆
Score

Math
DAY
16

/100

Date   /   /

Name

1 Divide.

(1) $12 \div 2 =$

(2) $15 \div 3 =$

(3) $28 \div 4 =$

(4) $30 \div 5 =$

(5) $24 \div 6 =$

(6) $42 \div 7 =$

(7) $56 \div 8 =$

(8) $81 \div 9 =$

(9) $8 \div 2 =$

(10) $24 \div 3 =$

(11) $45 \div 5 =$

(12) $48 \div 6 =$

(13) $21 \div 7 =$

(14) $16 \div 4 =$

(15) $35 \div 5 =$

(16) $18 \div 6 =$

(17) $63 \div 7 =$

(18) $10 \div 2 =$

(19) $27 \div 9 =$

(20) $54 \div 6 =$

(21) $72 \div 8 =$

(22) $30 \div 10 =$

(23) $54 \div 9 =$

(24) $9 \div 3 =$

(25) $32 \div 4 =$

Date   /   /

Name

① Read the passage. Use words from the passage to answer the   100 points for comp
questions below.

Rabbit ran into the bushes and began to beat his drum. Then Whale began to pull, and Elephant began to pull. In a minute the rope tightened until it was stretched as hard as a bar of iron.

"This is a very heavy cow," said Elephant, "but I'll pull her out." Bracing his huge feet in the earth, he gave a giant pull.

Soon Whale found himself sliding toward the land.

"No cow in the mud is going to beat me," Whale said.

He got so mad at the cow that he went head first, down to the bottom of the sea. That was a pull! Elephant was jerked off his feet, and came slipping and sliding toward the sea. He was very angry. Kneeling down on the beach, he twisted the rope around his trunk. He trumpeted and gave his hardest pull. Whale popped up out of the water and they bumped into each other. Then, they finally saw that each had hold of the same rope!

Rabbit rolled out of the bushes laughing. "You two couldn't rule all the animals, because you can't rule me!"

(1) Who bumps into each other?

Elephand and _whale_ bump into each other.

(2) What does Elephant twist the rope around?

Elephant twists the rope around his _trunk_.

(3) Where does Elephant kneel down?

Elephant kneels down on the _beach_.

(4) When does Whale go down to the bottom of the sea?

Whale goes down to the bottom of the sea after he begins _sliding_ toward the land.

(5) Why do Whale and Elephant bump into each other?

Whale and Elephant bump into each other because they each have hold of

the same _rope_.

(6) How does Rabbit signal Whale and Elephant to start pulling?

Rabbit signals Whale and Elephant by beating his _drum_.

Date / /

Name

1 Divide.

3 points per question

(1) $4 \div 2 = \boxed{2}$

(6) $6 \div 3 = \boxed{2}$

(11) $9 \div 4 = \boxed{2}R\boxed{1}$

(2) $5 \div 2 = \boxed{2}R\boxed{1}$

(7) $7 \div 3 = \boxed{2}R\boxed{1}$

(12) $17 \div 5 = \boxed{3}R\boxed{2}$

(3) $6 \div 2 = \boxed{3}$

(8) $8 \div 3 = \boxed{2}R\boxed{2}$

(13) $38 \div 6 = \boxed{6}R\boxed{2}$

(4) $7 \div 2 = \boxed{3}R\boxed{1}$

(9) $9 \div 3 = \boxed{3}$

(14) $25 \div 7 = \boxed{3}R\boxed{4}$

(5) $8 \div 2 = \boxed{4}$

(10) $10 \div 3 = \boxed{3}R\boxed{1}$

(15) $38 \div 8 = \boxed{4}R\boxed{6}$

2 Divide according to the example.

10 points for completion

$14 \div 3 = \boxed{7}R\boxed{1}$ → vertical form → $3\overline{)14}$ with $\boxed{7}R\boxed{1}$

3 Divide.

5 points per question

(1) $2\overline{)17}$ = $\boxed{8}R\boxed{1}$

(4) $4\overline{)25}$ = $6 r\boxed{1}$

(7) $7\overline{)35}$ = 5

(2) $2\overline{)18}$ = $\boxed{9}$

(5) $5\overline{)25}$ = 5

(8) $8\overline{)30}$ = $3 r\boxed{6}$

(3) $3\overline{)16}$ = $\boxed{5}R\boxed{1}$

(6) $6\overline{)35}$ = $5 r\boxed{5}$

(9) $9\overline{)54}$ = 6

Don't forget!
"R" means "remainder" or the amount left over after you've divided the numbers.

Reading
DAY
17

Chart the Passage
A Cooking Party 1

Date   /   /

Name

Level ⭐⭐

Score

/100

① Read the passage. Then answer the questions below.

Daphne's birthday was just around the corner. Her mom asked her, "What kind of party do you want to have?"

"I think I want a cooking party!" Daphne replied.

"That's a great idea! Let's plan it together," her mother said. "First we must choose a date, time, and location."

"Well, we already know we need to be in our kitchen," said Daphne.

"That's right," said her mom. "Then we must decide who to invite and brainstorm a menu. Then we can shop for supplies."

"Can we make chef hats, too? Then everyone can decorate theirs with their names." said Daphne.

"Another great idea! Very creative," said her mom.

(1) Use words from the passage above to complete the chart below.

75 points
for completio

How to prepare a cooking party
i. choose a _date_ , _time_ , and _location_
ii. decide who to _invite_
iii. brainstorm a _menu_
iv. _shop_ for supplies
v. make _chef_ hats

(2) Put a check (✓) next to the best title for the passage above.

25 points

(   ) Why I Wanted a Cooking Party   (   ) Going to a Party

(✓) How to Throw a Cooking Party   (   ) How to Prepare Cupcakes

Word Problems

Addition & Subtraction

Date / /

Name

Level ☆☆
Score

/100

Math
DAY
18

1 Read the word problem and write the number sentence below. Then 20 points per question answer the question.

(1) 353 people were in the amusement park. It started raining, so 95 people went home. How many people were left?

Ans. _____ people _____

(2) Mark's school has 186 boys and 189 girls. How many students go to Mark's school?

Ans. _____

(3) Mary read 157 pages yesterday. She read 256 pages today. How many pages did she read in all?

Ans. _____

(4) A restaurant had 355 onions this morning. The cooks used 287 of them. How many onions does the restaurant have left?

Ans. _____

(5) There are 187 boys and 178 girls at the park. Are there more boys or girls? How many more?

Ans. There are _____ more _____ .

Reading
DAY
18

Chart the Passage
A Cooking Party 2

Level ★★
Score

Date   /   /

Name

/100

1 Read the passage. Then answer the questions below.

Daphne thought about all her favorite treats: spaghetti and meatballs, watermelon, cookies and more. She decided she wanted to make homemade veggie pizza and ice cream sandwiches on her birthday. At the supermarket Daphne and her mom first got flour, salt and yeast for the pizza dough. Then they found the tomato sauce and shredded mozzarella cheese for the base. Then they picked out toppings: mushrooms, peppers, olives, eggplant, and onions. Last, they got ginger snap cookies and strawberry ice cream to make the ice cream sandwiches. Yum!

They also stopped at the arts and crafts store to get supplies for the chef hats. They bought white poster board, white tissue paper, tape, paper clips, and markers.

(1) Complete the chart below.

75 points for completion

Pizza layer	Ingredients
Dough	(1) flour (2) _____ (3) _____
Base	(4) tomato _____ (5) _____ mozzarella _____
(6)_____	(7) mushrooms (8) __peppers__ (9) _____ (10) _____ (11) _____

(2) Put a check (✓) next to the best title for the passage above.

25 points

(   ) Ingredients for Pancakes   (   ) Shopping for the Party

(   ) What Not to Buy for Pizza   (   ) How to Cook Pizza

Level ☆☆
Score

/100

Math
DAY
19

Date
/ /

Name

1 Read the word problem and write the number sentence below. Then answer the question. 20 points per question

(1) Lily put 2 apples on each dish. There are 5 dishes on the table. How many apples are there in all?

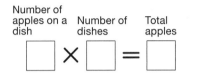

Number of apples on a dish Number of dishes Total apples

□ × □ = □

Ans. _____

(2) There are 3 vases in a room. Each vase has 3 flowers. How many flowers are there in all?

Ans. _____

(3) The teacher wants to give every student 2 pencils each. There are 9 students in the class. How many pencils will the teacher need?

Ans. _____

(4) In Jamal's homework group, there are 6 teams. Each team has 3 students. How many students are in Jamal's homework group?

Ans. _____

(5) Look at the picture below. How far is it from the red flag on the left to the blue flag on the right?

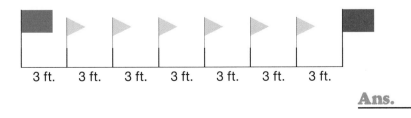

3 ft. 3 ft. 3 ft. 3 ft. 3 ft. 3 ft. 3 ft.

Ans. _____

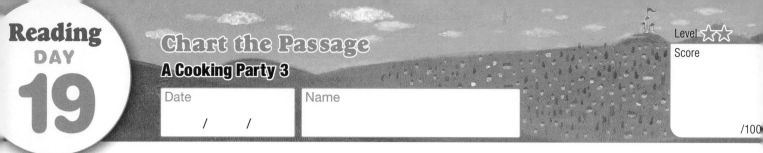

Reading
DAY
19

Chart the Passage
A Cooking Party 3

Level ⭐⭐

Score

/100

① Read the passage. Then answer the questions below.

First, make the dough: in a small bowl, dissolve yeast in warm water and add a dash of sugar. Meanwhile, in a mixing bowl add flour, salt, oil and spices. When yeast is bubbly, it's ready. Pour the yeast into the flour mixture and blend. Form your dough into a ball and place in a bowl that's been coated in oil. Let the dough rest 30 to 60 minutes. Roll out or stretch pizza dough into a pizza pan.

Then, add the base: spread sauce over stretched dough and sprinkle with cheese.

Last, put on toppings: choose your favorite vegetables as toppings and scatter them on the base.

Finally, ask a grown-up to bake your pizza at 450 degrees for 10 to 15 minutes.

(1) Complete the chart below.

75 points for completion

Steps	Directions
Make the dough	(1) _____ yeast in warm water and add sugar (2) mix together flour, _____, _____ and spices (3) pour the bubbly _____ into the flour mixture (4) form the _____ into a ball and let it rest (5) roll out the dough into a pizza _____
Add the base	(6) spread _____ over the stretched dough (7) _____ with cheese
(8) _____ _____	(9) choose your favorite _____ (10) _____ the toppings on the base

(2) Put a check (✓) next to the best title for the passage above.

25 points

(   ) Pizza Disaster   (   ) I Love Pizza

(   ) Veggie Pizza Recipe   (   ) Hot Ovens

Word Problems

Multiplication

Date
/ /

Name

Level ☆☆
Score

/100

Math
DAY
20

1 Read the word problem and write the number sentence below. Then answer the question. 20 points per question

(1) There are 8 benches in the train station. 4 people can sit on each bench. How many people can sit on the benches in all?

Ans. _____

(2) The art teacher gives 5 sheets of colored paper to each student. There are 7 students in art class today. How many sheets of colored paper will the art teacher need?

Ans. _____

(3) We had 6 pieces of tape that were 4 inches long each. Just for fun, we connected them end-to-end. How long was our new piece of tape?

| 4 in. | 4 in. | 4 in. | 4 in. | 4 in. | 4 in. |

Ans. _____

(4) The delivery man was making his rounds. He had 5 bags that each hold 6 packages. How many packages did he have?

Ans. _____

(5) The width of a pool is 5 meters. The length is four times the width. How long is the pool?

5 m

4 times the width

Ans. _____

Reading
DAY
20

Chart the Passage
Rising Mountains

Date / /

Name

Level ⭐⭐
Score

/100

① Read the passage. Then answer the questions below.

> Mountains are like wrinkles on the face of the earth. Over time, mountains rise up, collide, crack and fold. Most scientists say that a mountain is land that rises 1,000 feet (300 meters) or more above the surrounding area.
>
> The highest point on Earth is a folded mountain named Mount Everest. Mount Everest is in Nepal and is 29,035 feet (8,850 meters) high. The tallest mountain that is measured from top to bottom is Mauna Kea, which is in Hawaii. From the bottom, it is 33,474 feet (10,203 meters) tall but the mountain starts below the sea level, so it doesn't reach nearly as high above the earth's surface as Mount Everest. Mauna Kea is also a volcanic mountain, which forms when liquid rock from deep inside the Earth comes up through the ground and piles up. There are five main types of mountains: volcanic, dome, folded, plateau, and fault-block.

(1) Complete the chart below for Mount Everest.   **40** points for completion

Mountain name	Height	Location	Type of Mountain
Mount _____	_____ feet _____ meters	Nepal	folded

(2) Complete the chart below for Mauna Kea.   **40** points for completion

Mountain name	Height	Location	Type of Mountain
_____	_____ feet _____ meters	_____	_____

(3) Put a check (✓) next to the best title for the passage above.   **20** points

(   ) How to Climb a Mountain   (   ) Mauna Kea and Other Volcanoes

(   ) Facts About Mountains   (   ) Mountain Animals

Word Problems

Multiplication

Date	Name
/ /	

Level ☆☆

Score

/100

Math
DAY
21

1 Read the word problem and write the number sentence below. Then answer the question.

20 points per question

(1) Kim is going home for the holidays. She bought 6 boxes of candy as presents. Each box contains 6 candies. How many candies did she buy?

Ans. _____

(2) 1 week is equal to 7 days. How many days are in 4 weeks?

Ans. _____

(3) There are 5 children in your house. You want to give them each 7 pieces of fruit. How many pieces of fruit will you need?

Ans. _____

(4) The bookstore has 5 boxes which each hold 6 books. How many books does the bookstore have?

Ans. _____

(5) Steve was playing with bricks. He piled 8 bricks up, one on top of the other. If each brick was 8 centimeters thick, how tall was his pile?

Ans. _____

Sequencing
My Favorite T-shirt

Date / /

Name

Level ☆☆

Score

/100

① Read the title of the story. Then put the correct number under each picture so that the story is in the correct order.

100 points for comple

Making Tie-dyed T-shirts

(a)

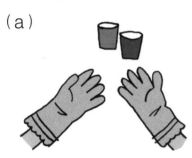

( ) Then I put on rubber gloves to work with the dye.

(b)

( ) I mixed the dye and water together in two tubs.

(c)

(|) First, I got all the supplies out.

(d)

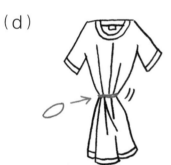

( ) Second, I put rubber bands around the shirt.

(e)

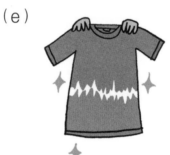

( ) I took off the rubber bands.
What a cool shirt!

(f)

( ) I put half of the white shirt in red dye for 5 minutes.

(g)

( ) I wrung out the extra dye from the shirt.

(h)

( ) I put the shirt in a plastic bag for a full day so the colors would set.

(i)

( ) Then I put the second half of the shirt in blue dye for 5 minutes.

1 Read the word problem and write the number sentence below. Then answer the question. 20 points per question

(1) The cafeteria has 9 boxes filled with juice bottles. If there are 8 bottles of juice in each box, how many bottles are there in all?

Ans. _____

(2) Each baseball team is made up of 9 players. If there are 6 teams waiting to play, how many players are there in all?

Ans. _____

(3) You used 4 sticks to make a square. If the sticks were 9 centimeters long, how long is the length around the square?

Ans. _____

(4) In Lora's class, there are 5 groups. Each group has 8 students. How many students are in her class?

Ans. _____

(5) You bought 7 sheets of drawing paper. 1 sheet costs 8¢. How much did you pay in all?

Ans. _____

Sequencing
On Your Mark, Get Set, Go!

Level ⭐⭐
Score
/100

Date   /   /

Name

1 Read the passage and answer the questions below.

> Paul was having his friends over, and he wanted to set up the biggest and hardest obstacle course ever. He searched all over his house for all the things he needed. Then he set up the course in his backyard and tested it himself. First, he raced through the ladder run. Then he swung across a lake made of pillows on his rope swing. Then he walked like a crab across the lawn to the golf tee. He putted a golf ball into his dog's house. Last, he hopped onto his bike and rode a lap around the whole thing to the finish line. Phew! He was tired but proud of his time. When his friends arrived he gave them each a number to pin onto their shirts. They took turns timing each other and trying to beat their own best times.

(1) Put the correct number next to each sentence so that the story is in the correct order.   80 points for complete

(   ) (a) Paul hopped through the ladder run.

(   ) (b) He rode to the finish line.

(   ) (c) They timed each other on the obstacle course.

(   ) (d) Paul swung on a rope swing.

(   ) (e) Paul got all the supplies.

(   ) (f) He set up each obstacle.

(   ) (g) Paul's friends arrived.

(   ) (h) He walked like a crab.

(2) Put a check (✓) next to the best title for the passage above.   20 points

(   ) Summer Fun

(   ) How to Climb a Mountain

(   ) How to Play Golf in Your Backyard

(   ) The Summer I Sprained My Ankle

(   ) Paul's Amazing Obstacle Course

Word Problems

Multiplication

Level ★★★

Score

/100

Math
DAY
23

Date / /

Name

1 Read the word problem and write the number sentence below. Then answer the question. 20 points per question

(1) There are 9 bags with 5 apples in each. How many total apples are there?

Ans. _____

(2) There are 7 bags with 8 stamps in each. How many total stamps are there?

Ans. _____

(3) Julia makes packets that have 4 candies each. If she gives 2 packets each to 7 people, how many candies is she giving away?

Number of candies per packet | Number of packets | Number of people | Total candies

☐ × ☐ × ☐ = ☐

Ans. _____

(4) Jeff makes packets that have 3 candies each. If he gives 3 packets each to 5 people, how many candies is he giving away?

Ans. _____

(5) The gardener gave each child 5 seeds to plant. If there are 48 children, how many seeds did the gardener give away?

Ans. _____

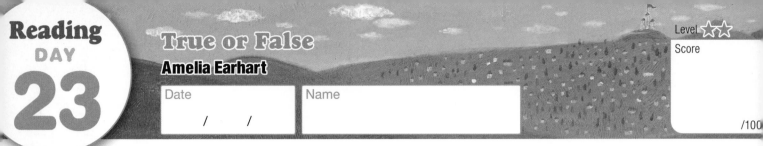

True or False
Amelia Earhart

Date / /

Name

① Read the passage. Then read the sentences below. Circle the "T" if the sentence is true, or correct. Circle the "F" if the sentence is false, or wrong. **10** points per question

Amelia Earhart is one of the world's most well-known pilots. She was the first woman to fly alone over the Atlantic Ocean. When Amelia was growing up, she moved around a lot with her family. After high school, she worked as an army nurse in Canada. When she was twenty-three years old, she began to learn how to fly although her family didn't want her to fly. Two years later she bought her first plane. In 1928, she became the first woman to fly across the Atlantic Ocean even though she was only a passenger. Amelia became more and more set on piloting a plane across the Atlantic alone, which she was able to do four years later. She finished the trip in record time—fourteen hours and fifty-six minutes. In 1937, Amelia tried to fly around the world. After finishing most of her trip, she and her plane vanished and were never found.

(1) Amelia's family wanted her to learn to fly. T (F)

(2) When Amelia was a girl, she moved around a lot. (T) F

(3) After college, she worked as an army nurse. T (F)

(4) Amelia began to learn to fly when she was twenty-three years old. (T) F

(5) Amelia was the second woman to fly alone over the Atlantic Ocean. T (F)

(6) Amelia bought her first plane when she was twenty-five years old. (T) F

(7) In 1928, Amelia became the first female pilot. T (F)

(8) Amelia set a record when she flew across the Atlantic. (T) F

(9) Amelia tried to fly around the world. (T) F

(10) Amelia's plane vanished while she tried to fly around the world. (T) F

Word Problems

Division

Date
/　/

Name

---

1 Read the word problem and write the number sentence below. Then answer the question.   20 points per question

(1) If 2 people want to share 8 candies equally, how many candies will each person get?

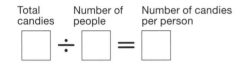

Total candies ÷ Number of people = Number of candies per person

Ans.

(2) You have 10 bananas. How many people will get bananas if you give them 2 bananas each?

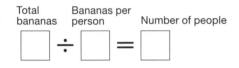

Total bananas ÷ Bananas per person = Number of people

Ans.

(3) We have 21 pencils for my group today. If the 3 of us share them equally, how many pencils will each of us get?

Ans.

(4) Your group in art class has 24 sheets of paper. If you want to give each person 3 sheets, how many people will get paper?

Ans.

(5) If you divide 45 centimeters of ribbon into 9 equal parts, how long would each of those parts be?

Ans.

True or False
Jellyfish

Date   /   /

Name

Level ⭐⭐

Score

/100

1 Read the passage. Then read the sentences below. Circle the "T" if the sentence is true, or correct. Circle the "F" if the sentence is false, or wrong.

10 points per question

Did you know that jellyfish have swum in the oceans for millions of years—even before dinosaurs walked the earth? Jellyfish can live almost anywhere wet—in cold or warm ocean water, along the coast, or in deep water. Jellyfish are mostly known for their squishy bodies and skinny, long arms. They can be bright colors like pink, yellow, or blue, or they can be clear. Some jellyfish even give off light! Jellyfish usually have a body that is shaped like a bell, and the opening is the mouth. Jellyfish swim around by squirting water out of their mouths to push them forward. Their arms can sting, stun, or even paralyze animals that they touch. Jellyfish don't attack people on purpose, but they can be dangerous if you accidentally touch a jellyfish arm. Some arms can reach as long as 100 feet (33 meters)! Watch out!

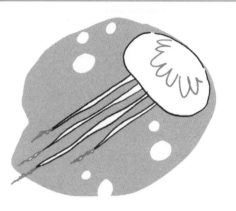

(1) Jellyfish are older than dinosaurs.   T   F

(2) Jellyfish can only live in warm ocean water.   T   F

(3) Most jellyfish have hard bodies and skinny, long arms.   T   F

(4) Jellyfish come in many different colors.   T   F

(5) All jellyfish give off light.   T   F

(6) Jellyfish don't attack people.   T   F

(7) Jellyfish have very short arms.   T   F

(8) Jellyfish move themselves forward by spitting
      water out of their mouths.   T   F

(9) Some jellyfish are shaped like a bassoon.   T   F

(10) Jellyfish can be dangerous.   T   F

Word Problems

Division with Remainders

Date
/ /

Name

Level ⭐⭐
Score
/100

Math
DAY
25

1 Read the word problem and write the number sentence below. Then answer the question. 20 points per question

(1) There are 20 cookies and 6 children. If they divide the cookies equally and everyone get 3 cookies. How many cookies remain?

 □ ÷ □ = □ R □

Ans. _____ cookies remain

(2) There are 30 cookies. If 7 children get 4 cookies each, how many cookies remain?

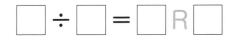 □ ÷ □ = □ R □

Ans. _____ cookies remain

(3) Your mother is making lunch. She divides 27 kiwis into 6 lunch bags evenly. How many kiwis are in each bag, and how many remain?

Ans. _____ kiwis in each bag, _____ kiwis remain

(4) Your brother is making lunch bags. He has 35 strawberries and puts 8 into each lunch bag. How many bags will he make, and how many strawberries will remain?

Ans. _____ bags, _____ strawberries remain

(5) June's mother has 50 roses, and she wants to put 8 in each vase. How many vases can she make? How many roses will she have left over?

Ans. _____ vases, _____ roses left over

Cause & Effect
The Vain Crow

Level ⭐⭐
Score

Date   /   /

Name

/10

1 Read the passage. Then answer the questions using words from passage.   20 points per question

> A crow named Cassius picked up some beautiful feathers left on the ground by the peacocks. He thought he would look better than the other crows if he stuck them into his own tail, so he did. In fact, Cassius thought he was now too fine to mix with the other crows. So he strutted off to the peacocks, and thought he'd be welcomed as one of them.
>
> The peacocks at once saw through his disguise. They disliked Cassius for being so vain. So they began to peck him, and soon he was stripped of all his borrowed feathers.
>
> Feeling a little naked and embarrassed, Cassius went sadly home. He wanted to join his old crow friends as if nothing had happened. But they remembered how he had mocked them. They chased him away and would have nothing to do with him.
>
> "If you had been happy," said one crow, "to remain as nature made you, instead of trying to be what you are not, you would not have been refused by the others or disliked by your equals."

(1) Why did Cassius stick the peacock feathers in his tail?

He thought he would _____ _____ than the other crows.

(2) What happened when Cassius went to be with the peacocks?

The peacocks saw through his _____.

(3) Why did the peacocks begin to peck Cassius?

The peacocks began to peck Cassius because they disliked him for

being so _____.

2 Complete the chart with words from the passage above.   40 points for completi

Cause	Result or Effect
Because the peacocks didn't accept Cassius.	Cassius wanted to _____ his old crow friends as if _____ had happened.
Because Cassius mocked his friends, the crows.	The crows _____ him away and would have _____ to do with him.

Large Numbers

Level ★★

Score

/100

Math
DAY
26

Date
/   /

Name

1 Answer the following questions about the number "37,456,812."

8 points per question

(1) Fill the words in each box below.

place	place	place	ten-thousands place	thousands place	hundreds place	tens place	ones place
3	7	4	5	6	8	1	2

(2) Reading from the thousands place to the left, we see the ⬚

place, the ⬚ place, the ⬚ place and the

⬚ place.

(3) The 7 in 37,456,812 is in the ⬚ place.

(4) The 4 in 37,456,812 is in the ⬚ place.

(5) The 3 in 37,456,812 is in the ⬚ place.

2 Write the correct number in each box.

8 points per question

(1) **60,000** is the number you get from adding ⬚ ten-thousands.

(2) **260,000** is the number you get from adding ⬚ ten-thousands.

(3) **260,000** is the number you get from adding ⬚ thousands.

(4) **2,600,000** is the number you get from adding ⬚ ten-thousands.

(5) **2,600,000** is the number you get from adding ⬚ thousands.

3 Compare the numbers below. Write < or > in the boxes.

5 points per question

(1) 5,000 ⬚ 3,000

(2) 690,000 ⬚ 67,000

(3) 101,101 ⬚ 101,110

(4) 1,801,012 ⬚ 1,800,901

Don't forget!
> means "is greater than"
< means "is less than"

Reading

DAY

26

Cause & Effect

Level ☆☆

Score

/10

Date / /

Name

1 Match the cause with the effect.

10 points per question

Cause	Effect

(1) Jessica was thirsty...

(2) The elephant got scared...

(3) The sun was setting...

(4) Mary hated drawing...

(5) The beach was dirty...

(6) The ants found a crumb...

(7) I broke a glass by accident...

(8) She was tired...

(9) The dog was hungry...

(10) It was raining...

ⓐ ... so instead she played guitar.

ⓑ ... so my mom got a broom and carefully swept it up.

ⓒ ...so he ran away.

ⓓ ...so they carried it back to their anthill.

ⓔ ... so he dug up his bone.

ⓕ ...so she got a glass of juice.

ⓖ ... so she took a nap.

ⓗ ...so soon it would be dark outside.

ⓘ ... so we couldn't have a picnic.

ⓙ ...so we got trash bags to clean it up.

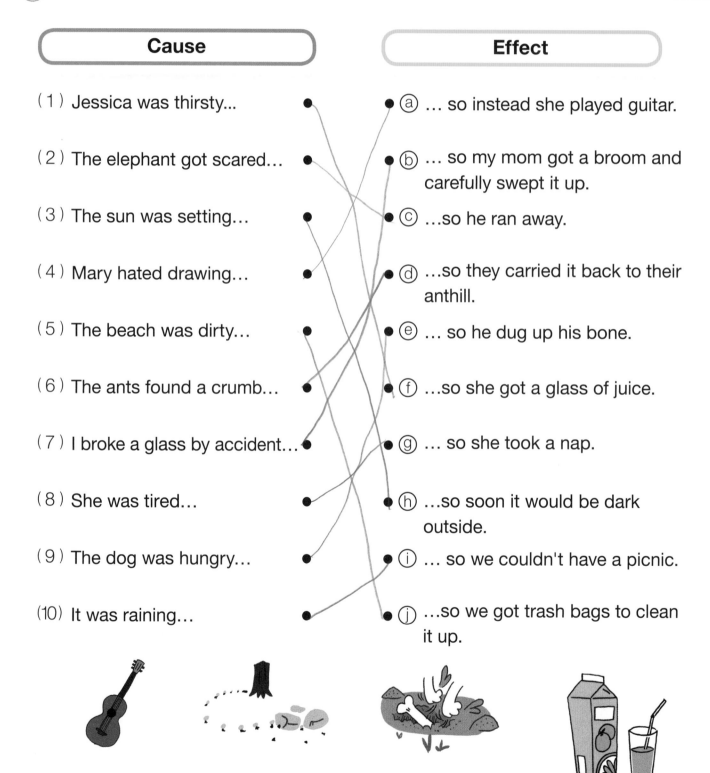

Fractions

Level ☆

Score

/100

Math
DAY
27

Date   /   /

Name

1 How many parts of the whole figure are shaded?   8 points per question

(1)

$\left(\dfrac{1}{2} \right)$

(3)

$\left(\dfrac{1}{4} \right)$

(5)

$\left(\dfrac{1}{5} \right)$

(7)

$\left(\dfrac{1}{5} \right)$

(2)

$\left(\dfrac{1}{3} \right)$

(4)

$\left(\dfrac{1}{4} \right)$

(6)

$\left(\dfrac{1}{5} \right)$

(8)

$\left(\dfrac{1}{6} \right)$

2 Each piece of tape is 1 foot long and has been divided into equal   6 points per question
parts. How many parts of the whole piece of tape are shaded in each
figure below?

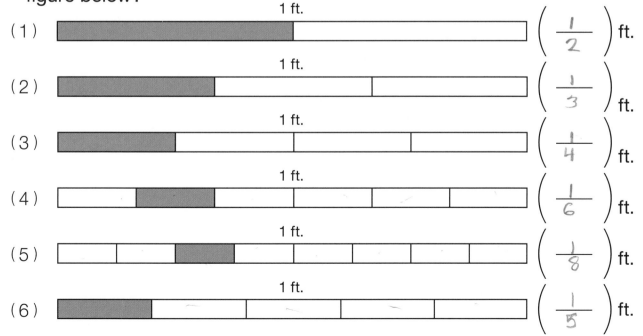

(1)   1 ft. $\left(\dfrac{1}{2} \right)$ ft.

(2)   1 ft. $\left(\dfrac{1}{3} \right)$ ft.

(3)   1 ft. $\left(\dfrac{1}{4} \right)$ ft.

(4)   1 ft. $\left(\dfrac{1}{6} \right)$ ft.

(5)   1 ft. $\left(\dfrac{1}{8} \right)$ ft.

(6)   1 ft. $\left(\dfrac{1}{5} \right)$ ft.

Main Idea
The Butter Cow

Date / /

Name

① Read the passage. Then answer the questions below.

Every year, the Iowa State Fair celebrates butter in many special ways—from showing people how to make butter to a butter cookie competition. But the greatest attraction is the famous Butter Cow.

The Butter Cow starts with a frame made out of wood, metal, wire and steel. The frame is put in a cooler that is about forty degrees. Then about six hundred pounds of pure Iowa butter is added layer upon layer. Once enough butter is on the frame, the artist carves a life-size cow that is about 5.5 feet (1.7 meters) high and 8 feet (2.4 meters) long.

John Karl Daniels made the first Iowa State Fair butter cow in 1911. Since then, there have only been four other people who have sculpted the Butter Cow. In 1960, Norma "Duffy" Lyon was the first woman to sculpt the Butter Cow, and she did it every year for forty-six years!

The Iowa State Fair Butter Cow turned one hundred years old in 2011. That year, the fair made one hundred cows from all kinds of objects like cans, flowers and sand, and placed them throughout the fairgrounds.

(1) What is the main idea of the second paragraph? Put a check (✓) next to the correct sentence below.   30 points

(   ) The Iowa State Fair celebrates butter.

(   ) The Butter Cow is made by sculpting butter on a frame.

(   ) The Butter Cow must be made in a 40 degree cooler.

(2) What is the main idea of the last paragraph? Put a check (✓) next to the correct sentence below.   30 points

(   ) Every year the Iowa State Fair has the Butter Cow.

(   ) Cows can be made from sand, too.

(   ) The Iowa State Fair celebrated 100 years of the Butter Cow in 2011.

(3) Put a check (✓) next to the best title for the passage above.   40 points

(   ) The Iowa State Fair's Butter Cow

(   ) How to Make a Butter Cow

(   ) New Uses for Butter

Don't forget! A **main idea** is a sentence that states the most important information in a passage or paragraph.

Fractions

Level ⭐⭐

Score

/100

Math
DAY
28

Date / /

Name

1 Write each fraction.

10 points per question

(1) 1 foot is divided into 5 equal parts, and 4 parts are shaded. $\left(\dfrac{4}{5}\right)$

(2) 1 foot is divided into 5 equal parts, and 3 parts are shaded. $\left(\dfrac{3}{5}\right)$

(3) 1 foot is divided into 7 equal parts, and 3 parts are shaded. $\left(\dfrac{3}{7}\right)$

(4) 1 foot is divided into 7 equal parts, and 5 parts are shaded. $\left(\dfrac{5}{7}\right)$

(5) 1 foot is divided into 8 equal parts, and 3 parts are shaded. $\left(\dfrac{3}{8}\right)$

2 Write the number of shaded parts as a fraction.

10 points per question

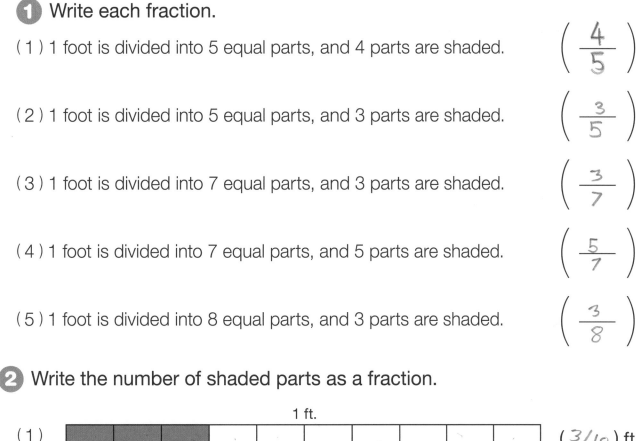

(1) (3/10) ft.

(2) (5/10) ft.

(3) (9/10) ft.

(4) (4/10) ft.

(5) (7/10) ft.

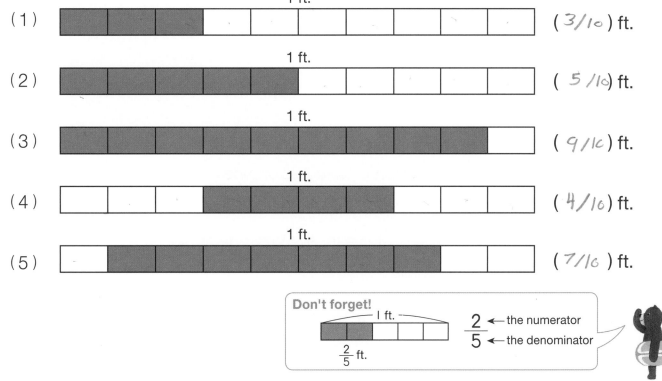

Don't forget!

1 ft.

$\dfrac{2}{5}$ ft.

$\dfrac{2}{5}$ ← the numerator
← the denominator

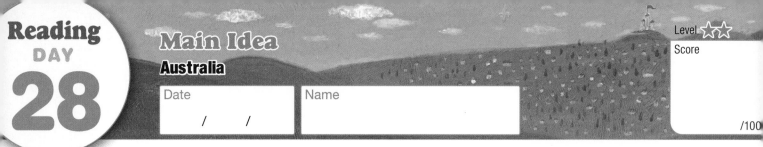

Reading
DAY
28

Main Idea
Australia

Date / /

Name

Level ⭐⭐
Score

/100

(1) **Read the passage. Then answer the questions below.**

> Australia is a special place. It is the only country that is also a continent. A continent is a large piece of land on the globe, for example North America or Asia. Australia has many different kinds of land—from desert to rainforest. Australia also has many rare animals and plants that live there and nowhere else, like the kangaroo.
>
> One of the most famous areas of Australia is the outback. It has the country's hottest weather and very few animals are able to live there. There are large deserts with very little water and almost no plants. Native people, called Aborigines, have learned to live in this difficult climate.
>
> However, most people live on the edges of the country near the coast where the weather is more mild. Australians love to play sports in Australia's warm weather and great outdoors. Australians swim, surf, sail, and play soccer. Australians have even invented their own type of football.

(1) What is the main idea of the first paragraph? Put a check (✓) next to the correct sentence below.   30 points

   (   ) Australia has a large desert.

   (   ) The weather in Australia changes often.

   (✓) Australia is a unique place.

(2) What is the main idea of the second paragraph? Put a check (✓) next to the correct sentence below.   30 points

   (   ) Australia has rare animals.

   (   ) Native people of Australia are called Aborigines.

   (✓) The outback is a tough place to live.

(3) Put a check (✓) next to the best title for the passage above.   40 points

   (   ) How to Travel to Australia

   (✓) A Guide to Australia

   (   ) Australian Sports

Length

Level ☆☆
Score

/100

Math
DAY
29

Date / /

Name

5,280
+ 80
5,360

5,280
+ 700
5,980

10,560
+ 100
10,660

1 Convert the measurements below.

5 points per question

(1) 1 mi. = [5,280] ft.

(2) 1 mi. 3 ft. = [5,283] ft.

(3) 1 mi. 80 ft. = [5,360] ft.

(4) 1 mi. 700 ft. = [5,980] ft.

(5) 2 mi. 100 ft. = [10,660] ft.

(6) 2 mi. 20 ft. = [10,580] ft.

(7) 3 mi. 10 ft. = [   ] ft.

(8) 3 mi. 220 ft. = [   ] ft.

(9) 5,280 ft. = [1] mi.

(10) 5,285 ft. = [1] mi. [5] ft.

(11) 6,000 ft. = [   ] mi. [   ] ft.

(12) 8,000 ft. = [   ] mi. [   ] ft.

(13) 10,560 ft. = [   ] mi.

(14) 11,000 ft. = [   ] mi. [   ] ft.

(15) 14,000 ft. = [   ] mi. [   ] ft.

(16) 15,600 ft. = [   ] mi. [   ] ft.

2 Order the following lengths from longest to shortest with the numbers 1 through 4.

10 points per question

(1)   1 mi.   5,370 ft.   1 mi. 50 ft.   5,300 ft.

(   )   (   )   (   )   (   )

(2)   10,550 ft.   2 mi.   1 mi. 600 ft.   16,000 ft.

(   )   (   )   (   )   (   )

Don't forget!
5,280 feet (ft.) is equal to 1 mile.
(2 mi. = 10,560 ft., 3 mi. = 15,840 ft., ...)

Reading
DAY
29

Characters
The Robin and the Pitcher

Date
/   /

Name

Level ☆☆
Score

/100

(1) Read the passage. Then answer the questions below.

20 points per question

> A red robin, whose throat was dry from singing, saw a large pitcher in the distance. Happily, he flew to it, but found that it held only a little water, and even that was too near the bottom to be reached.
>
> The robin bent down and stretched his neck, but he had no luck. Next he tried to tip the pitcher, thinking that he would at least be able to catch some of the water as it fell out. But he was not strong enough to move it at all. He even asked his friend the worm to try to push the pitcher with him, but the worm was really no help.
>
> He walked round and round the edge of the pitcher and sang a tune to help himself think. Then he saw some pebbles lying nearby. He had an idea! He picked up the pebbles and dropped them one by one into the pitcher. After many pebbles, he managed at last to raise the water up to the very top, and took a good, long drink.

(1) Put a check (✓) next to the words that describe the robin.

(   ) strange   (   ) sad   (   ) dirty

(   ) thirsty   (   ) strong   (   ) musical

(   ) clever   (   ) red   (   ) mean

(2) What did the robin want?

The robin wanted some _____.

(3) Who tried to help the robin?

The _____ tried to help the robin.

> **Don't forget!**
> A **character** is an individual (sometimes a person or animal or even an object) in a story, novel or play. The author gives characters actions, thoughts, and speech.

(4) What did the robin do to help himself think?

The robin _____ a tune to help him think.

(5) Does the robin achieve his goal?

The robin _____ achieve his goal.

Length

Level ★★
Score

Math
DAY
30

Date   /   /

Name

/100

1 1 centimeter (cm) is equal to 10 millimeters (mm). How far is each box ⟨30 points for completion⟩ from the left side of the ruler?

| mm | | cm | mm | | | | |

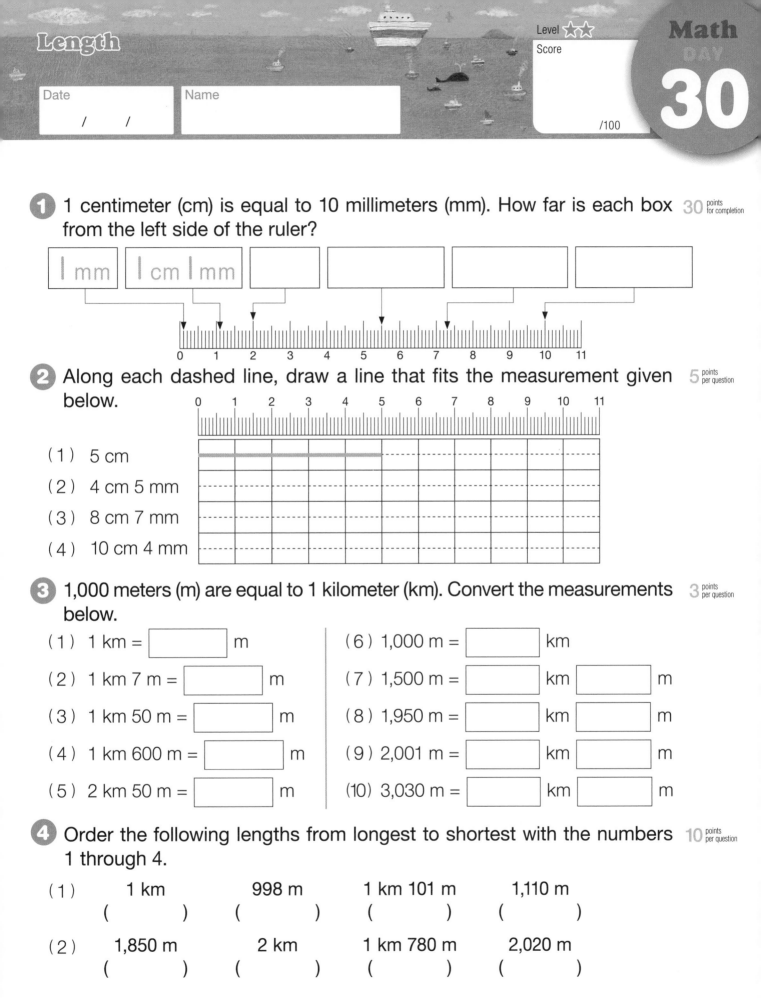

2 Along each dashed line, draw a line that fits the measurement given ⟨5 points per question⟩ below.

(1) 5 cm

(2) 4 cm 5 mm

(3) 8 cm 7 mm

(4) 10 cm 4 mm

3 1,000 meters (m) are equal to 1 kilometer (km). Convert the measurements ⟨3 points per question⟩ below.

(1) 1 km = ⬚ m

(2) 1 km 7 m = ⬚ m

(3) 1 km 50 m = ⬚ m

(4) 1 km 600 m = ⬚ m

(5) 2 km 50 m = ⬚ m

(6) 1,000 m = ⬚ km

(7) 1,500 m = ⬚ km ⬚ m

(8) 1,950 m = ⬚ km ⬚ m

(9) 2,001 m = ⬚ km ⬚ m

(10) 3,030 m = ⬚ km ⬚ m

4 Order the following lengths from longest to shortest with the numbers ⟨10 points per question⟩ 1 through 4.

(1)   1 km   998 m   1 km 101 m   1,110 m
        (   )   (   )   (   )   (   )

(2)   1,850 m   2 km   1 km 780 m   2,020 m
        (   )   (   )   (   )   (   )

Reading
DAY
30

Characters

Why the Bear has a Stumpy Tail

Level ☆☆

Score

Date / /

Name

/100

① Read the passage. Then answer the questions below.

> One winter day the brown bear saw the fox who was slinking along with some fish he had stolen.
>
> "Hey! Stop a minute! Where did you get those?" demanded the bear.
>
> "Oh, my. Well, I've been out fishing and caught them," the fox lied.
>
> So the bear wanted to learn to fish, too, and asked the fox to tell him how he was to set about it.
>
> "Oh, it is quite easy," answered the fox, "and simple to learn. You've only got to go upon the ice, and cut a hole and stick your tail down through it, and hold it there as long as you can.
>
> Don't pay attention if it smarts a little; that's when the fish bite. The longer you hold it there, the more fish you'll get; and then all at once, out with it! You must pull sideways, and give a strong pull, too."
>
> Well, the bear did as the fox said, and though he felt very cold, and his tail smarted very much, he kept it in a long, long time, He wanted all the fish in the pond! But at last his tail was frozen in the ice, though of course he did not know that. When he yanked it out, his tail snapped right off, and that's why the bear walks around with a stumpy tail to this day!

(1) Which character had stolen some fish?

The _____ had stolen some fish.

(2) What did the bear want?

The bear wanted all the _____ in the _____.

(3) Put a check (✓) next to the words that describe the bear.

( ) hungry   ( ) tricky   ( ) slimy

( ) thirsty   ( ) greedy   ( ) brown

( ) clever   ( ) tired   ( ) skinny

(4) Put a check (✓) next to the words that describe the fox.

( ) sneaky   ( ) unhappy   ( ) talented

( ) heavy   ( ) strong   ( ) mean

( ) clever   ( ) hot   ( ) sly

Area

1 The figures below are made of squares that have 1-inch sides. What is the area of each figure?   10 points per question

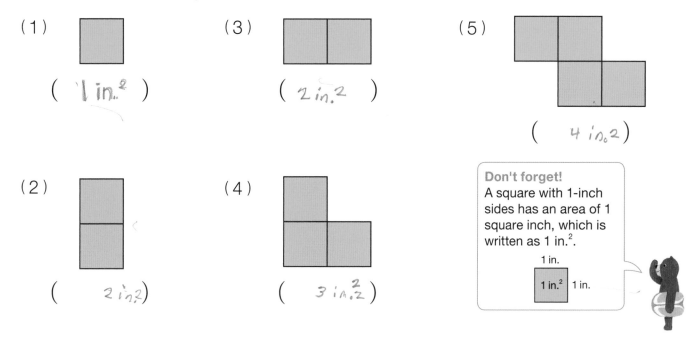

(1) (1 in.²)

(3) (2 in.²)

(5) (4 in.²)

(2) (2 in.²)

(4) (3 in.²)

Don't forget!
A square with 1-inch sides has an area of 1 square inch, which is written as 1 in.².

1 in.

1 in.² 1 in.

2 The figures below are made of squares that have 1-centimeter sides. What is the area of each figure?   10 points per question

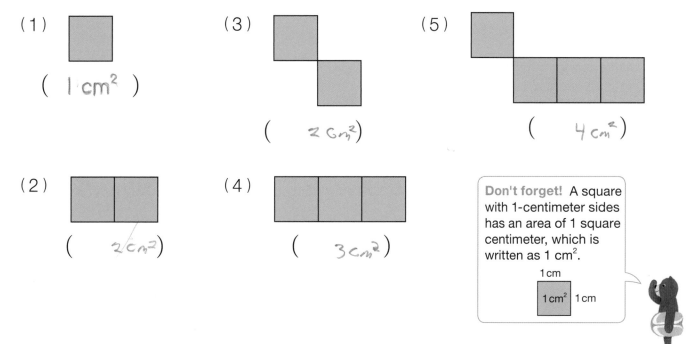

(1) (1 cm²)

(3) (2 cm²)

(5) (4 cm²)

(2) (2 cm²)

(4) (3 cm²)

Don't forget! A square with 1-centimeter sides has an area of 1 square centimeter, which is written as 1 cm².

1 cm

1 cm² 1 cm

Reading
DAY
31

Reading Comprehension
The Story of Peter Pan 1

Level ★★★
Score

Date   /   /

Name

/100

1 Read the passage below. Then answer the questions.

20 points per question

> Once upon a time there were three children named Wendy, John, and Michael, who lived with their father and mother in London. One evening the father and mother were invited to a party, and the mother, after lighting the dim lamp in the nursery and kissing the children good-night, went away. Later that evening, a little boy climbed in through the window. His name was Peter Pan. He was a curious little fellow, very conceited, very forgetful, and yet very lovable. The most remarkable thing about him was that he never grew up. There came flitting in through the window with him his fairy, whose name was Tinker Bell.
>
> Peter Pan woke all the children up, and after he had sprinkled fairy dust on their shoulders, he took them away to Neverland, where he lived with a family of lost boys.

(1) In what city does this story take place?

The story takes place in _London_ .

(2) In which characters' bedroom does the story begin?

The story begins in _wendy_ , _John_ **and** _micheal_ **'s bedroom.**

(3) Does the story start during the day or during the night?

The story starts during the _euning_ .

(4) Where does Peter Pan take the children?

Peter Pan takes the children to _Neverland_ .

(5) What is the setting of the first paragraph?

The setting of the first paragraph is in the evening, in the children's

house **in the city of** _London_ .

> **Don't forget!** The **setting** is the background (specifically the time and place) of a story.

Weight

Date / /

Name

/100

1 Read each scale. Then write the weight below.

5 points per question

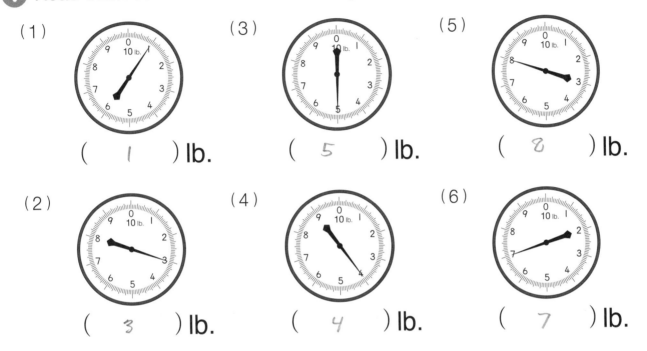

(1) (1) lb.

(3) (5) lb.

(5) (8) lb.

(2) (3) lb.

(4) (4) lb.

(6) (7) lb.

2 16 ounces (oz.) are equal to 1 pound (lb.). Convert the measurements below.

5 points per question

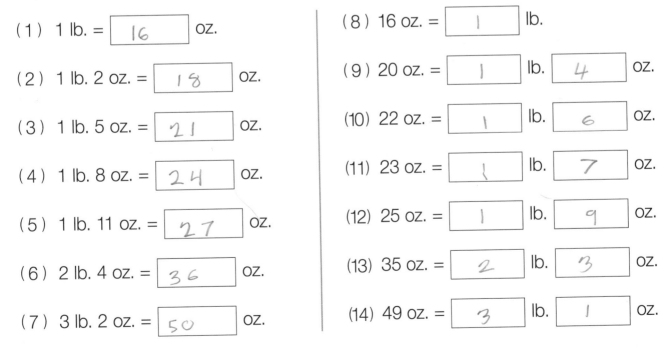

(1) 1 lb. = 16 oz.

(2) 1 lb. 2 oz. = 18 oz.

(3) 1 lb. 5 oz. = 21 oz.

(4) 1 lb. 8 oz. = 24 oz.

(5) 1 lb. 11 oz. = 27 oz.

(6) 2 lb. 4 oz. = 36 oz.

(7) 3 lb. 2 oz. = 50 oz.

(8) 16 oz. = 1 lb.

(9) 20 oz. = 1 lb. 4 oz.

(10) 22 oz. = 1 lb. 6 oz.

(11) 23 oz. = 1 lb. 7 oz.

(12) 25 oz. = 1 lb. 9 oz.

(13) 35 oz. = 2 lb. 3 oz.

(14) 49 oz. = 3 lb. 1 oz.

Reading
DAY
32

Reading Comprehension
The Story of Peter Pan 2

Level ★★★
Score

Date / /

Name

/1

① Read the passage below. Then answer the questions.   20 points per que.

> Tinker Bell was jealous of the little girl Wendy, and she hurried ahead of Peter Pan and persuaded the lost boys that Wendy was a bird who might do them harm, and so one of the boys shot her with his bow and arrow.
>
> When Peter Pan came and found Wendy lying lifeless upon the ground in the woods he was very angry, but he was also very quick-witted. So he told the boys that if they build a house around Wendy, she would be better. So they hurried to collect everything they had out of which they could make a house.

(1) What are the main events of this passage?

  (a) Tinker Bell persuades the _____ that Wendy might do them harm.

  (b) One of the boys shoots Wendy with his _____.

  (c) Peter Pan finds _____ lying lifeless in the woods.

  (d) Peter Pan _____ the boys to _____ a house around Wendy.

  (e) The boys _____ everything to make a house.

(2) Why is Peter Pan angry?

  Peter Pan is angry because one of the boys _____ Wendy.

(3) Why do the boys build a house?

  The boys build a house to make Wendy _____.

② Read the passage below. Then answer the questions.   40 points for comple

> When the house was done, Peter Pan took John's hat for the chimney. The little house was so pleased to have such a capital chimney that smoke at once began to rise through the hat. All that night, Peter Pan walked up and down in front of Wendy's house to watch over her and keep her from danger while she slept.

What are the main events of this passage?

  (a) Peter Pan takes John's _____ to make the _____.

  (b) The house likes its chimney and puffs _____.

  (c) Peter _____ up and down in front of Wendy's house.

Don't forget!
The **plot** is the main events of a story.

Weight

Level ☆☆
Score

/100

Math
DAY
33

Date

/ /

Name

1 1,000 grams (g) are equal to 1 kilogram (kg). Convert the measurements below. 5 points per question

(1) 1 kg = ☐ g

(2) 1 kg 30 g = ☐ g

(3) 1 kg 300 g = ☐ g

(4) 2 kg 10 g = ☐ g

(5) 2 kg 200 g = ☐ g

(6) 3 kg 60 g = ☐ g

(7) 3 kg 578 g = ☐ g

(8) 1,000 g = ☐ kg

(9) 1,008 g = ☐ kg ☐ g

(10) 1,110 g = ☐ kg ☐ g

(11) 2,002 g = ☐ kg ☐ g

(12) 2,201 g = ☐ kg ☐ g

(14) 3,005 g = ☐ kg ☐ g

(15) 3,205 g = ☐ kg ☐ g

2 Read each scale. Then write the weight below. 5 points per question

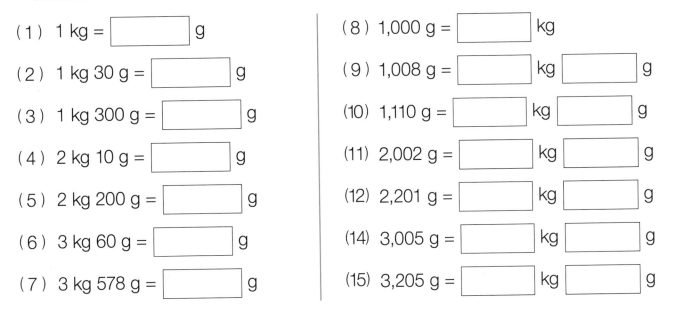

(1)

(400 g   )

(3)

(   )

(5)

(   )

(2)

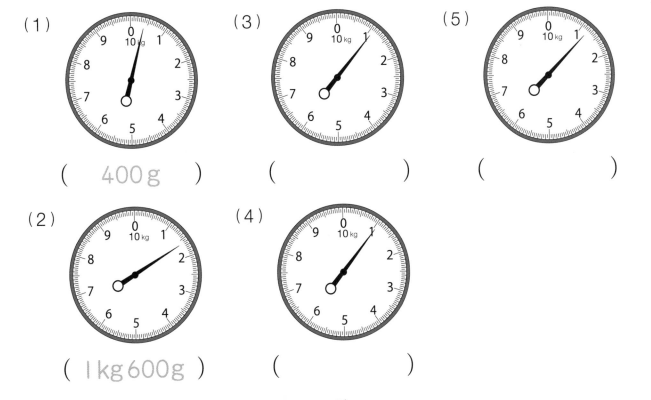

(1 kg 600 g)

(4)

(   )

Reading
DAY
33

Reading Comprehension
The Story of Peter Pan 3

Date / /

Name

Level ★★★
Score

/100

① Read the passage below. Then answer the questions.

25 points per question

One day, Wendy and her brothers realized that they had been away so long that perhaps their mother had forgotten them and shut the window of the nursery so that they could not get back. They decided to hurry home. When they reached home, Peter Pan was before them, and he closed the window so that they could not return. But when he heard the children's mother singing such a sad song inside, his heart was made tender and he opened the window. The children crept back safely into their mother's arms. Wendy's mother invited Peter Pan to stay and be her child, but Peter was so afraid that he would have to go to school and grow up and be a man that he went back to his home in fairyland. Wendy promised to go once a year and stay a few days with Peter Pan to clean the house and mend his clothes. Let us picture them in the little house that was built for Wendy, which the fairies had put up in the branches of a pine-tree. The birds are singing in their nests and in the branches, and far below the clouds you can see the land and the sea.

（1） Why did the children want to return home?

The children wanted to return home because they were afraid that their

mother had _____ them.

（2） Why did Peter Pan close the window?

Peter Pan closed the window because he didn't want the children

to _____ to their home.

（3） Why did Peter Pan return to fairyland?

Peter Pan returned to fairyland because he was _____ he would
have to grow up.

（4） Put a check (✓) next to the phrases or words that could be the story's theme.

( ) growing up   ( ) learning to fly   ( ) the love of music

( ) friendship   ( ) mothering   ( ) travel

Don't forget! The **theme** is the subject of the story, or an idea that forms the story.

Weight

Date / /

Name

1 1,000 milligrams (mg) are equal to 1 gram (g). Convert the measurements below.

5 points per question

(1) 1 g = [   ] mg

(2) 3 g = [   ] mg

(3) 7 g = [   ] mg

(4) 1,000 mg = [   ] g

(5) 4,000 mg = [   ] g

(6) 5,000 mg = [   ] g

2 Order the following weights from heaviest to lightest with the numbers 1 through 4.

10 points per question

(1)   1 lb.   17 oz.   1 lb. 2 oz.   15 oz.
       (   )   (   )   (   )   (   )

(2)   29 oz.   2 lb.   2 lb. 2 oz.   33 oz.
       (   )   (   )   (   )   (   )

(3)   1 kg   1,100 g   1 kg 90 g   900 g
       (   )   (   )   (   )   (   )

(4)   2 kg 300 g   2,090 g   2 kg 290 g   2,110 g
       (   )   (   )   (   )   (   )

(5)   3 kg 850 g   4 kg   3,900 g   3 kg 99 g
       (   )   (   )   (   )   (   )

(6)   1 g   1,101 mg   999 mg   1,090 mg
       (   )   (   )   (   )   (   )

(7)   1 mg   1 kg   1 g   100 mg
       (   )   (   )   (   )   (   )

Reading
DAY
34

Reading Comprehension
Simile

Level ★★★

Score

Date / /

Name

/100

① Read the sentences below. Circle the things that are being compared in the simile.   10 points per question

(1) The girl was as graceful as a swan.

(2) The boat drifted like a cloud.

(3) Jeff's cannonball into the pool was like a bomb.

(4) At the talent show, Robin was cool as a cucumber.

(5) When I looked down from the plane's window, the farmland looked like a quilt.

(6) My brother dances like a robot.

(7) The woman's red hair was like fire.

(8) My backpack feels as deep as the ocean.

② Choose words from the box below to finish the simile. Then circle the words that compare the two things.   4 points per question

| oven   molasses   mountain   fountain   lightning |

(1) The car drove (as) fast (as) _____.

(2) The summer day was hot (like) an _____.

(3) The dog was drooling like a _____.

(4) The football player was solid like a _____.

(5) My chubby cat was as slow as _____.

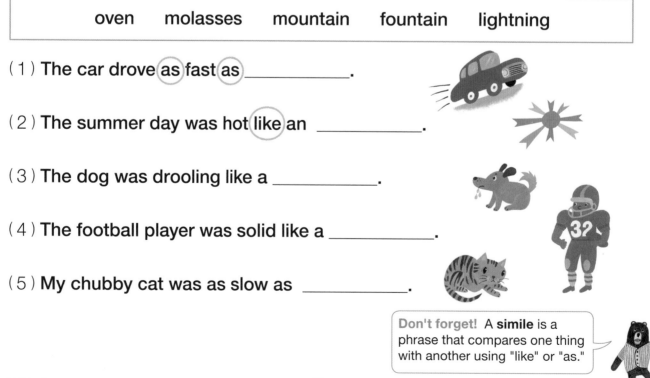

> **Don't forget!** A **simile** is a phrase that compares one thing with another using "like" or "as."

Capacity

Level ☆☆

Score

/100

Math
DAY
35

Date   /   /

Name

1 Look at the water in each cup below. Then order the cups from most water to least water with the numbers 1 through 3.   13 points per question

(1)   (   )   (   )   (   )

(2)   (   )   (   )   (   )

(3)   (   )   (   )   (   )

(4)   (   )   (   )   (   )

2 What is the total amount of water shown in each problem below?   8 points per question

(1) 1 pint

(1 pt.)

(2) 1 pint   1 pint

(   )

(3) 1 pint 1 pint

1 pint 1 pint   (   )

(4) 1 quart

(1 qt.)

(5) 1 quart   1 quart

(   )

(6) 1 quart   1 quart

1 quart   (   )

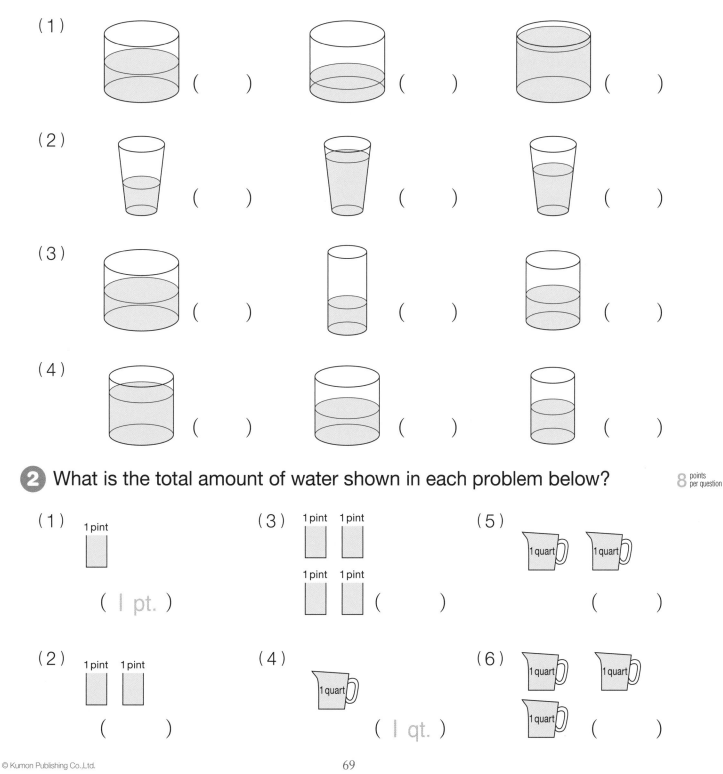

Reading Comprehension
Metaphor

Date   /   /   Name

1 Read the sentences below. Circle the things that are being compared in the metaphor.

10 points per question

(1) Her eyes were bright jewels.

(2) The dog's tail was a drummer.

(3) The goalie was a wall.

(4) The bully was a toad.

(5) His computer was a dinosaur.

(6) My mom is an angel.

(7) Her room was a train-wreck.

(8) By the time I got to it, the ice cream was a pool.

2 Choose words from the box below to finish the metaphor.

4 points per question

| apples   lockbox   spaghetti   music   statue |

(1) Our teacher's brain is a _lockbox_.

(2) Your cheeks are red _____.

(3) After the race, my legs were _____.

(4) The lady's voice was _____.

(5) The scared mouse is a _____.

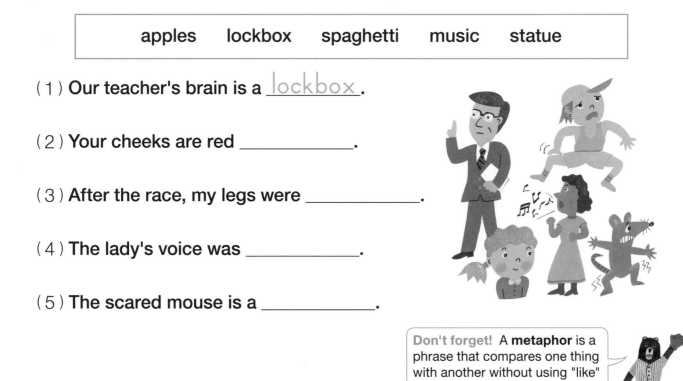

Don't forget! A **metaphor** is a phrase that compares one thing with another without using "like" or "as."

Capacity

1 Convert the measurements below.

4 points per question

(1) 1 qt. = [   ] pt.

(2) 1 qt. 1 pt. = [   ] pt.

(3) 2 qt. 1 pt. = [   ] pt.

(4) 1 gal. = [   ] qt.

(5) 1 gal. 3 qt. = [   ] qt.

(6) 1 gal. = [   ] pt.

(7) 1 gal. 2 qt. = [   ] pt.

(8) 2 pt. = [   ] qt.

(9) 4 pt. = [   ] qt.

(10) 7 pt. = [   ] qt. [   ] pt.

(11) 5 qt. = [   ] gal. [   ] qt.

(12) 9 qt. = [   ] gal. [   ] qt.

(13) 8 pt. = [   ] gal.

(14) 10 pt. = [   ] gal. [   ] qt.

2 What is the total amount of water shown in each problem below?

11 points per question

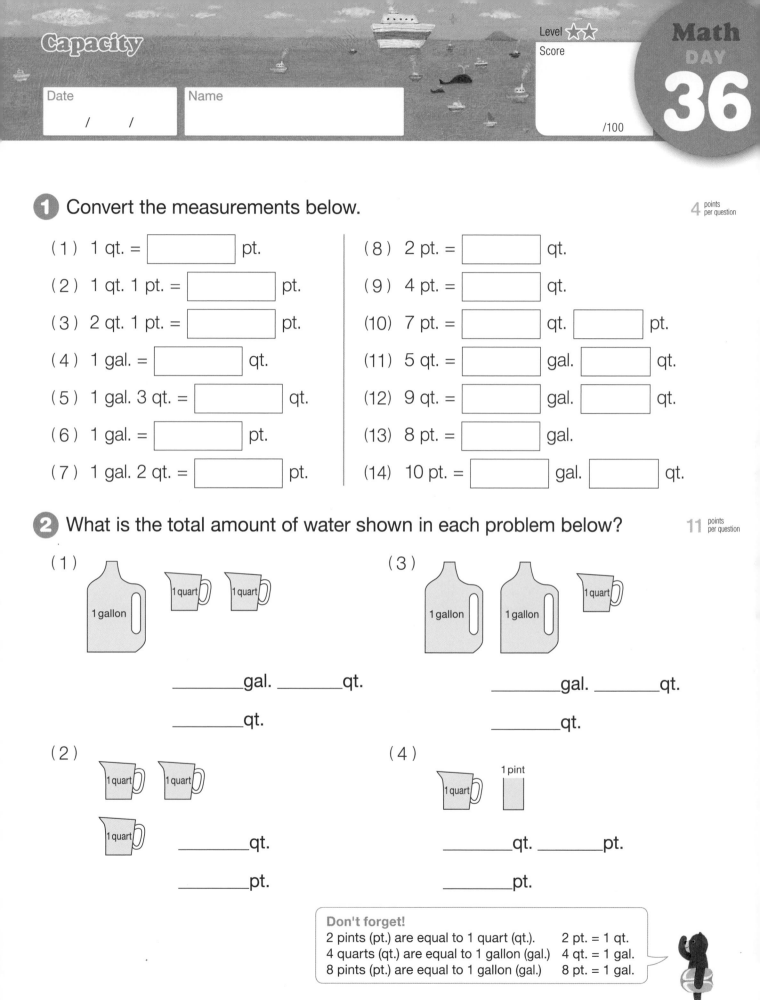

(1)

_____gal. _____qt.

_____qt.

(2)

_____qt.

_____pt.

(3)

_____gal. _____qt.

_____qt.

(4)

_____qt. _____pt.

_____pt.

Don't forget!
2 pints (pt.) are equal to 1 quart (qt.).   2 pt. = 1 qt.
4 quarts (qt.) are equal to 1 gallon (gal.)   4 qt. = 1 gal.
8 pints (pt.) are equal to 1 gallon (gal.)   8 pt. = 1 gal.

Reading Comprehension
Poetry

Date   /   /

Name

① Read the poem below. Then answer the questions.

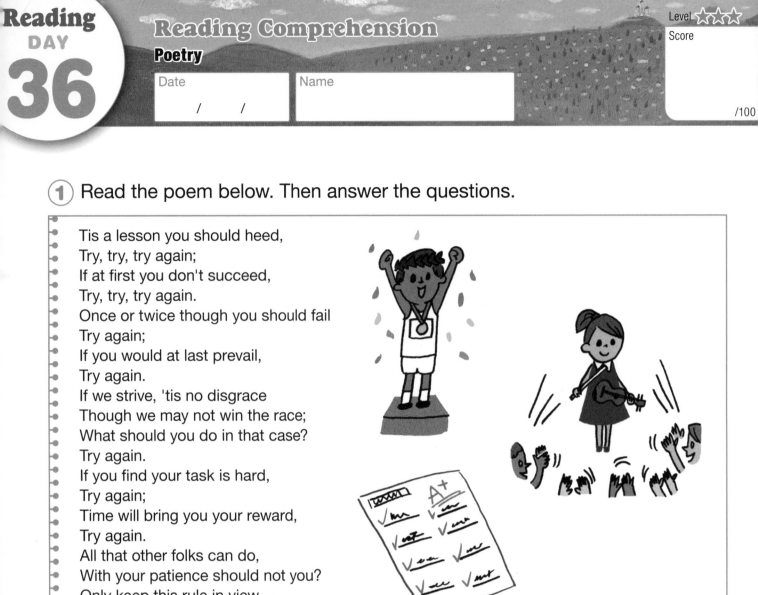

Tis a lesson you should heed,
Try, try, try again;
If at first you don't succeed,
Try, try, try again.
Once or twice though you should fail
Try again;
If you would at last prevail,
Try again.
If we strive, 'tis no disgrace
Though we may not win the race;
What should you do in that case?
Try again.
If you find your task is hard,
Try again;
Time will bring you your reward,
Try again.
All that other folks can do,
With your patience should not you?
Only keep this rule in view—
Try again.

(1) Put a check (✓) next to the phrases that describe the poem's main idea.　　60 points for completion

(✓) hard work will pay off 　　　(✓) do your best

(   ) being lazy is fun 　　　　　　(✓) work for your goals

(   ) don't break the rules 　　　　(   ) goals are too hard

(2) What are the first two rhyming words?　　20 points for completion

The first two rhyming words are ___heed___ **and** ___Succeed___ .

(3) Put a check (✓) next to the best title for the poem above.　　20 points

(   ) Success Is the Best

(✓) Try Again

(   ) Failing Isn't Fun

Capacity

Level ☆☆

Score

Math
DAY
37

/100

Date   /   /

Name

1 1,000 milliliters (mL) are equal to 1 liter (L). Convert the measurements below.   5 points per question

(1) 1 L = $\boxed{1,000}$ mL

(2) 3 L = $\boxed{3,000}$ mL

(3) 5 L = $\boxed{5,000}$ mL

(4) 1,000 mL = $\boxed{1}$ L

(5) 4,000 mL = $\boxed{4}$ L

(6) 5,000 mL = $\boxed{5}$ L

2 How much water is in each container below?   10 points per question

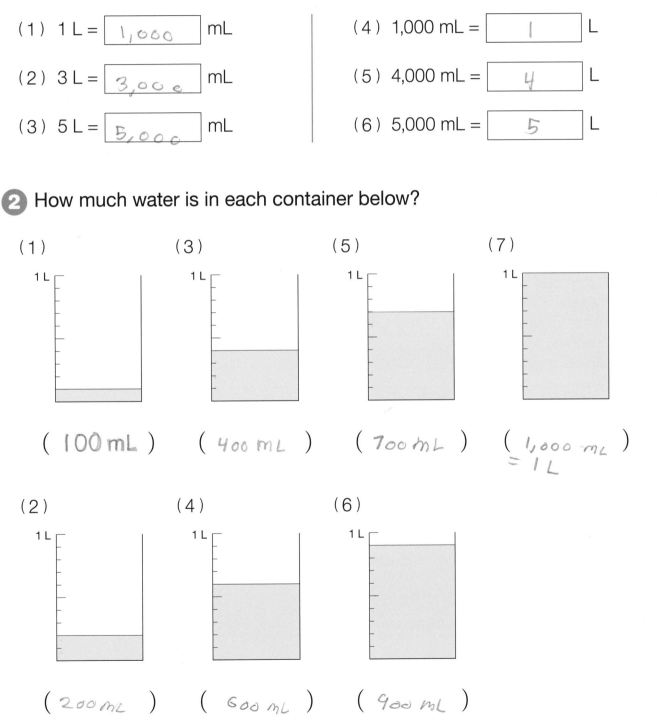

(1)

1 L

(100 mL)

(3)

1 L

(400 mL)

(5)

1 L

(700 mL)

(7)

1 L

(1,000 mL)
= 1 L

(2)

1 L

(200 mL)

(4)

1 L

(600 mL)

(6)

1 L

(900 mL)

Reading Comprehension
Poetry

Date   /   /

Name

① Read the poem below, "If I Ever See" by Lydia Maria Child. Then answer the questions.   25 points per question

If ever I see,
On bush or tree,
Young birds in their pretty nest,
I must not in play,
Steal the birds away,
To grieve their mother's breast.
My mother, I know,
Would sorrow so,
Should I be stolen away;
So I'll speak to the birds
In my softest words,
Nor hurt them in my play.
And when they can fly
In the bright blue sky,
They'll warble a song to me;
And then if I'm sad
It will make me glad
To think they are happy and free.

(1) Put a check (✓) next to the phrases that describe the poem's main idea.

(   ) catching birds is fun   (✓) stealing birds is bad

(✓) wild birds belong in nature   (   ) birds are good pets

(✓) wild birds should be free   (   ) young birds can't fly

(2) What are the first two rhyming words?

The first two rhyming words are ___see___ **and** ___tree___.

(3) What are the second two rhyming words?

The second pair of rhyming words are ___play___ **and** ___away___.

(4) What will make the speaker in the poem glad?

The speaker in the poem will be glad to think the birds are ___happy___

and ___free___.

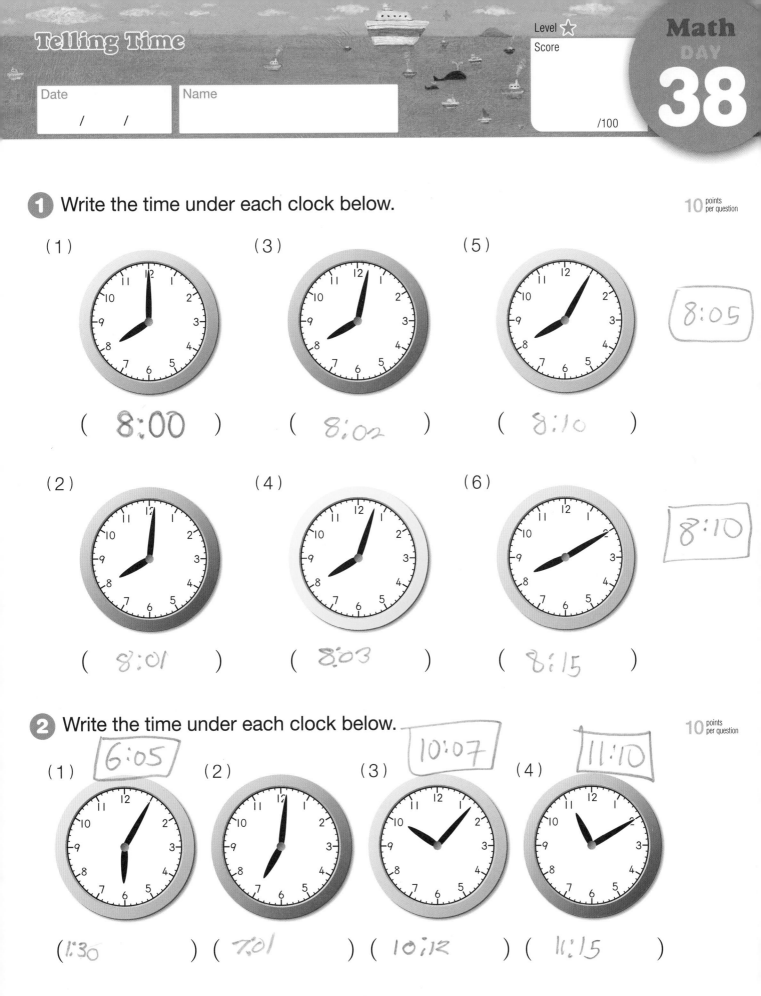

Telling Time

Level ☆
Score

/100

Math
DAY
38

1 Write the time under each clock below.

10 points per question

(1)
(8:00)

(3)
(8:02)

(5)
8:05
(8:10)

(2)
(8:01)

(4)
(8:03)

(6)
8:10
(8:15)

2 Write the time under each clock below.

10 points per question

(1) 6:05
(1:30)

(2)
(7:01)

(3) 10:07
(10:12)

(4) 11:10
(1:15)

Reading
DAY
38

Reading Comprehension
The Dragonfly of Lookout Mountain 1

Level ★★★
Score

Date / /

Name

/10

① Read the passage from *The Dragonfly of Lookout Mountain* by Judy Hatch. Then answer the questions below. 20 points for comple

> The new dragonfly nymph dropped into the water. He sank down to the muddy bottom of the pond.
>
> The nymph was an expert hunter. He had six sturdy legs and a ferocious hooked mouthpart which he could shoot out from under his chin. He caught and ate anything which could not get away, and he grew quickly. When he grew too big for his skin, he shed it and formed a new one the next size up.

(1) Complete the chart below.

Characteristics of the dragonfly nymph	
(a) an _____ hunter	(c) ferocious _____ mouthpart
(b) _____ sturdy legs	(d) grows _____

② Read the passage. Then answer the questions below. 20 points per questi

> While the spring passed, he grew quickly. He ate insects, then tadpoles, and even small fish. He shed his skin many times. By summer he was two inches long and the largest and most powerful insect in the pond.
>
> A turtle and a trout also lived in his part of the pond. The nymph knew better than to let them see him.
>
> Most of the time he hid in the soft bottom weeds. If they came too close, he swam quickly by squirting water out behind himself, and squeezed into cracks in rocks where he knew they couldn't find him.

(1) Who shed his skin many times?

The _____ shed his skin many times.

(2) What did the nymph eat?

The nymph ate _____, _____, and even small _____.

(3) Where did the nymph hide?

The nymph hid in the soft bottom _____ and in the cracks in _____.

(4) When was the nymph two inches long?

The nymph was two inches long by _____.

Telling Time

Level ☆☆
Score

/100

Math
DAY
39

Done! 9/04/20

1 Write the time under each clock below.

10 points per question

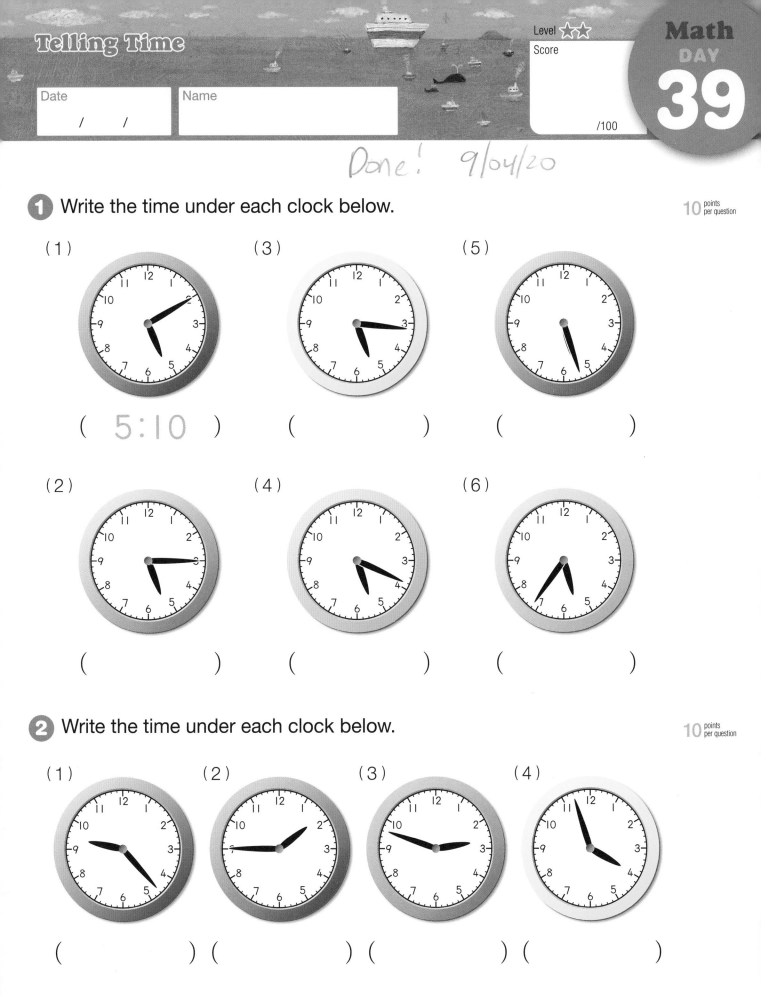

(1)

(5:10)

(3)

(   )

(5)

(   )

(2)

(   )

(4)

(   )

(6)

(   )

2 Write the time under each clock below.

10 points per question

(1)

(   )

(2)

(   )

(3)

(   )

(4)

(   )

Reading
DAY
39

Reading Comprehension
The Dragonfly of Lookout Mountain 2

Level ★★★
Score

Date   /   /

Name

/100

① Read the passage. Then answer the questions below.

> By being brave and clever, the nymph had survived in the pond for almost a year. He knew about the dangers of living in the water, but he had not yet learned that danger could come from above the water, too.
>
> When a raccoon reached down and snatched him from the bottom, he didn't know what was happening until it was too late. He grabbed some weeds and held on. He kicked his legs and struggled.
>
> The raccoon shook the weeds away. He turned the nymph over in his hands and sniffed him. He moved his hands toward his mouth. The insect felt hot breath washing over him. He saw the mouth and jagged teeth coming closer.
>
> With no time to escape, the nymph did the only thing he could. He grabbed hard with his legs, lashed out his mouthpart, and bit deep into the raccoon's nose.
>
> The animal jumped back and knocked the nymph loose. The insect tumbled through the air and landed on his back in the pond.

(1) Number the sentences to match the order they occur in the passage.   50 points for comple

(3) The nymph held onto some weeds.

(5) The nymph bit the raccoon's nose.

(1) A raccoon reached down into the pond.

(6) The raccoon sprang back, and the nymph fell into the pond.

(4) The raccoon turned the nymph over and over in his hands.

(2) The raccoon grabbed the nymph.

(2) Put a check (✓) next to the best answer to each question.   25 points per questi

(a) Why did the raccoon pick up the nymph?

(   ) To ask the nymph a question   (   ) To joke with the nymph

(✓) To eat the nymph   (   ) To help the nymph

(b) How did the nymph escape?

(   ) By distracting the raccoon   (✓) By biting the raccoon

(   ) By feeding the raccoon   (   ) By swimming fast

78

Triangles & Quadrilaterals

Level ☆☆
Score

Math
DAY
40

/100

Date
/ /

Name

1 Sort the shapes into the categories below.

20 points per question

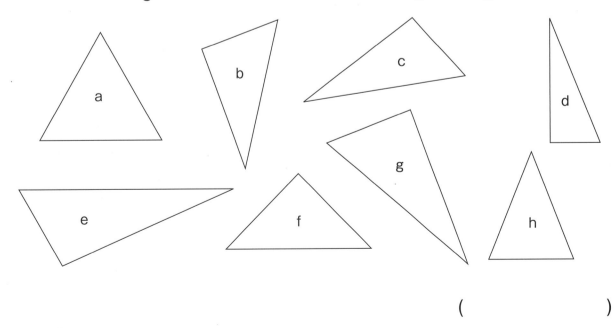

(1) Square   (   )
(2) Rectangle   (   )
(3) Not a square and not a rectangle   (   )

2 Look at the triangles and list the letters for the right triangles below.

40 points for completion

(   )

Reading
DAY
40

Reading Comprehension
The Dragonfly of Lookout Mountain 3

Date / /

Name

Level ⭐⭐⭐
Score

/10

1 Read the passage below. Then answer the questions.   25 points per question.

Zigzagging up the road, he came to the fire tower. The lookout saw the sparkle of his wings and bronze green body against the darkening clouds.

"Have you come to visit me?" she asked.

The dragonfly, as though answering, waved his legs and flew by her face. He looked carefully at her with his thousands of hexagonal eyes.

She stayed very still.

He saw that part of her was soft and flowing like weeds. Her eyes were round, like the trout's, but not so terrible. Her mouth was dark in opening and light in closing like the turtle's, but soft. Her hands were like the raccoon's, but she had round claws and she did not try to grab him. She lived by herself in a huge tall nest and if she had wings, they were well hidden.

(1) Put a check (✓) next to the adjectives that describe the dragonfly.

(✓) sparkly   (✓) green   (✓) bronze

( ) dull   (✓) winged   ( ) still

(2) Put a check (✓) next to the phrases that describe the lookout.

(✓) round eyes   ( ) green body   ( ) sparkly

(✓) lives alone   ( ) winged   (✓) still

( ) sharp nails   (✓) long hair   ( ) jumping

(3) What is the main idea of the last paragraph? Put a check (✓) next to the correct sentence below.

(✓) The dragonfly observes the lookout.

( ) The lookout tries to catch the dragonfly.

( ) The dragonfly visits an old friend.

(4) Put a check (✓) next to the best title for the passage above.

( ) Being a Lookout

( ) Dragonflies and Monsters

(✓) When the Dragonfly Meets the Lookout

80

Boxes

Level ☆☆
Score

/100

Math
DAY
41

Date
/ /

Name

1 If you constructed boxes by folding the figures on the left, what kinds 20 points per question
of boxes would you make? Connect the figures on the left to the
correct boxes on the right.

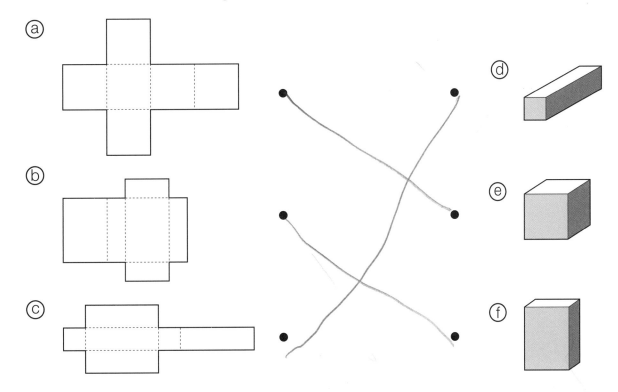

ⓐ

ⓑ

ⓒ

ⓓ

ⓔ

ⓕ

2 The boxes pictured here are opened. However, face A is missing from 20 points per question
each picture. Add face A to the figures below.

(1)

(2)

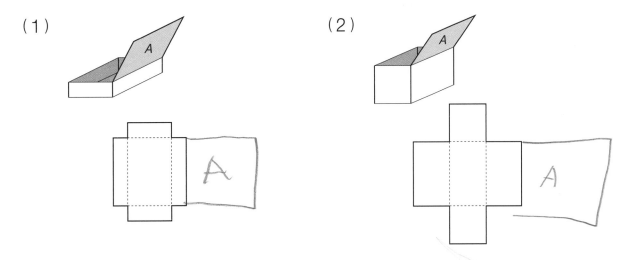

Reading
DAY
41

Reading Comprehension
The Dragonfly of Lookout Mountain 4

Date   /   /

Name

Level ★★★
Score

/10

1 Read the passage. Then answer the questions below.   30 points per question

> A dark cloud rumbled overhead and a swirl of cold air washed by, pushing dust plumes down the road. The rope slapped the flagpole. The lookout's hair whipped across her face. While she watched, the dragonfly's wings yanked back. He jerked sideways and then was ripped away upward in the wind.
>   "Oh no!" she exclaimed, watching wide-eyed. "Come back!"
>   But she knew he couldn't. The wind was much too strong.

(1) Put a check (✓) next to the adjectives that describe the setting.

(   ) sunny   (   ) windy   (   ) bronze

(   ) dusty   (   ) snowing   (   ) wild

(2) Complete the chart according to the passage above.

Cause	Effect
the _____	The lookout's hair whipped across her face.
	The _____ slapped the flagpole.
	The dragonfly could not come back to the _____.

2 Read the passage. Then answer the question below.   40 points

> It began to hail.
>   The dragonfly saw the deadly hailstones. He knew that the blow of one against his fragile skin would crush him, and even the gentle kiss of any brushing by could shatter his transparent wings.

(1) Which of the following is the dragonfly most likely to do next? Put a check (✓) next to the best answer.

(   ) The dragonfly will look for food.

(   ) The dragonfly will look for shelter.

(   ) The dragonfly will fly to the pond.

Tables

Level ☆☆

Score

/100

Date / /

Name

Math DAY 42

1 Complete the exercises below based on the charts.

25 points per question

Class A	
Fruit	Number of Students
Apples	10
Oranges	8
Melons	11
Peaches	4
Grapes	2
Pears	3

Class B	
Fruit	Number of Students
Apples	9
Oranges	10
Melons	8
Peaches	5
Grapes	3
Pears	2

Class C	
Fruit	Number of Students
Apples	11
Oranges	7
Melons	10
Peaches	6
Grapes	4
Pears	1

(1) Calculate the total for each type of fruit and write it in the appropriate box in the table on the right.

(2) Which class has the most students?

(class c)

(3) Which fruit is favored by the most students?

(apples)

(4) What does the number in j represent?

(the total of students)

Class Favorite Fruit				
Fruit	Class A	Class B	Class C	Total
Apples	10	9	11	d 30
Oranges	8	10	7	e 25
Melons	11	8	10	f 29
Peaches	4	5	6	g 15
Grapes	2	3	4	h 9
Pears	3	2	1	i 6
Total	a 38	b 37	c 39	j 114

Reading Comprehension
The Dragonfly of Lookout Mountain 5

Date   /   /

Name

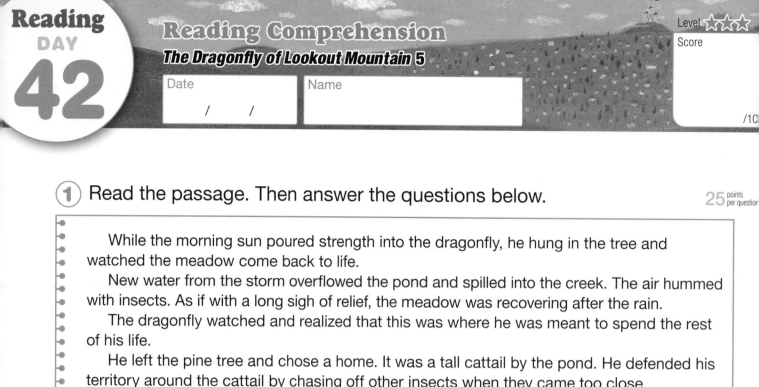

① Read the passage. Then answer the questions below.

25 points per question

> While the morning sun poured strength into the dragonfly, he hung in the tree and watched the meadow come back to life.
> New water from the storm overflowed the pond and spilled into the creek. The air hummed with insects. As if with a long sigh of relief, the meadow was recovering after the rain.
> The dragonfly watched and realized that this was where he was meant to spend the rest of his life.
> He left the pine tree and chose a home. It was a tall cattail by the pond. He defended his territory around the cattail by chasing off other insects when they came too close.

(1) What are the main events of this passage?

(a) After the storm, the dragonfly _____ in the tree.

(b) Water _____ the pond and _____ into the creek.

(c) The air _____ with insects.

(d) The dragonfly _____ a _____ as his home.

(e) The dragonfly _____ his territory.

(2) Put a check (✓) next to the phrases or words that could be the story's theme.

(   ) home   (   ) friendship   (   ) nature's wonder
(   ) survival   (   ) war   (   ) love

(3) Which of the following is the dragonfly most likely to do next? Put a check (✓) next to the best answer.

(   ) The dragonfly will look for food.

(   ) The dragonfly will leave his cattail for vacation.

(   ) The dragonfly will choose a new home close to the raccoon.

(4) Put a check (✓) next to the best title for the passage above.

(   ) Dragonfly Finds a Home

(   ) The Lookout and the Dragonfly's New Home

(   ) Why Cattails are Good Homes

Review

Level ☆☆
Score

/100

Math
DAY
43

Date   /   /

Name

1 Multiply.   5 points per question

(1) $4 \times 2 = 8$ (5) $4 \times 7 = 28$ (9) $9 \times 5 = 45$

(2) $6 \times 5 = 30$ (6) $8 \times 10 = 80$ (10) $6 \times 7 = 42$

(3) $8 \times 1 = 8$ (7) $3 \times 3 = 9$ (11) $2 \times 8 = 16$

(4) $2 \times 6 = 12$ (8) $7 \times 5 = 35$ (12) $3 \times 2 = 6$

2 There are 6 children and you want to give them 8 pieces of candy each.   10 points
How many pieces of candy will you need?

6×8

$6 \times 8 = 48$

Ans. 48 you will need 48 pieces of candy

3 Each baseball team is made up of 9 players. If there are 7 teams waiting to   15 points
play, how many players are there in all?

$9 \times 7 = ?$

$9 \times 7 = 63$

Ans. There are 63 players

4 There are 24 cookies and 7 children. If they divide the cookies equally, how   15 points
many cookies does each child get? How many cookies are left over?

$24 \div 7 =$ $3 \times 7 = 21$

Ans. each child gets 3 cookies 3 are
Left over.

85

Review
Sea Stars

Date   /   /

Name

1 Complete the passage using the vocabulary words defined below.   60 points for completi

> Did you know that starfish are actually not fish at all? Although most people know them as starfish, many scientists are trying to change the name to sea star. Fish have backbones but sea stars do not. Animals without backbones, like sea stars, are called invertebrates. Animals with backbones, like fish, are called _____.
>
> Sea stars are very _____ creatures. Sea stars can _____ a new arm if one is lost. Most have five arms, but some sea stars can grow as many as fifty arms. Their arms have _____ and suckers that help them grip the ground and move.
>
> Most sea stars also have the remarkable ability to _____ food outside their bodies. Sea stars use their feet to open clams. The stomach _____ from the mouth and goes inside the clam shell. The stomach then _____ the prey and goes back into the body when it is done.

vertebrates: have a backbone
unique: rare, or one of a kind
pincers: jaws used for gripping or pinching
regenerate: regrow
consume: eat or drink
emerges: moves out or comes into view
envelops: wraps up, covers, or surrounds

2 Use the passage above to answer the questions below.   10 points per questio

(1) Who wants to change the name of the starfish to sea star?

_____ **want to change the name of the starfish to sea star.**

(2) What don't sea stars have?

Sea stars don't have _____.

(3) How do sea stars move?

Sea stars have _____ and _____ that help them _____ the ground.

(4) Why don't scientists want to use the name starfish?

Because these animals are not actually _____.

Review

Level ☆☆

Score

/100

Math
DAY
44

Date   /   /

Name

1 Divide.   4 points per question

(1) $6 \div 2 =$

(2) $6 \div 3 =$

(3) $8 \div 4 =$

(4) $8 \div 3 =$ R

(5) $15 \div 5 =$

(6) $20 \div 4 =$

(7) $36 \div 6 =$

(8) $42 \div 7 =$

(9) $56 \div 8 =$

(10) $50 \div 6 =$ R

(11) $50 \div 7 =$ R

(12) $63 \div 9 =$

(13) $70 \div 8 =$ R

(14) $48 \div 5 =$ R

(15) $24 \div 3 =$

2 We have 24 pencils for my group today. If the 6 of us share them equally,   20 points
how many pencils will we each get?

Ans. _____

3 There are 45 cookies. If 7 children get 6 cookies each, how many cookies   20 points
are left over?

Ans. _____

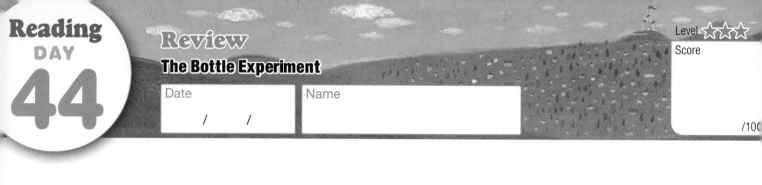

Reading
DAY
44

Review
The Bottle Experiment

Date	Name
/ /	

Level ★★★
Score

/100

1 Read the passage. Then answer the questions below.

Stephanie and her little brother Nick were walking home from school. Stephanie was blowing into her half-full water bottle and tooting a tune to pass the time. Nick asked, "What makes the sound in your bottle?"

"Well, I could tell you," Stephanie said, "but I think I'll show you too!"

When they got home, Stephanie set up the experiment. She went into the kitchen and got five identical plastic bottles out of the recycling bin. Then she asked Nick to fill them up with different amounts of water. Then she arranged the bottles in order from most full to least full.

"Now blow across the top of each bottle and listen to the sounds," Stephanie instructed.

"They're different!" Nick said after he had blown up and down the row of bottles until he was out of breath.

"That's right! When you blow across the top, you make the air inside shake and shiver. The bottles with more air sound lower than the bottles with more water."

(1) Complete the chart below.   50 points for completion

The Bottle Music Experiment
i. Get five _____.
ii. Fill them with different amounts of _____.
iii. Arrange the bottles from most _____ to least _____.
iv. _____ across the top and _____ to the sounds.

(2) Complete the chart with words from the passage above.   50 points for completion

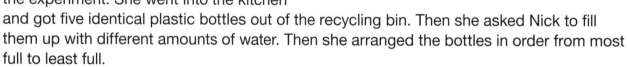

Cause	Effect
more air in the bottle	The sound is _____.
more _____ in the bottle	The sound is higher.

Review

Level ⭐⭐
Score

Math
DAY
45

Date
/   /

Name

/100

1 Write each fraction.

10 points per question

(1) 1 foot is divided into 5 equal parts, and 2 parts are shaded.   (——)

(2) 1 foot is divided into 7 equal parts, and 5 parts are shaded.   (——)

(3) 1 foot is divided into 9 equal parts, and 4 parts are shaded.   (——)

2 Convert the measurements below.

5 points per question

(1) 1 mi.   = ☐ ft.

(2) 1 mi. 3 ft.= ☐ ft.

(3) 5,290 ft. = ☐ mi. ☐ ft.

(4) 6,000 ft. = ☐ mi. ☐ ft.

(5) 1 km 101 m = ☐ m.

(6) 2,010 m = ☐ km ☐ m

3 Order the following groups from heaviest to lightest with the numbers 1 through 4.

10 points per question

(1)   1 lb.   18 oz.   1 lb. 1 oz.   14 oz.
       (   )   (   )   (   )   (   )

(2)   1,010 g   999 g   1 kg   1 kg 9 g
       (   )   (   )   (   )   (   )

(3)   1,009 mg   1 g   1,100 mg   999 mg
       (   )   (   )   (   )   (   )

4 What is the total amount of water shown in each problem below?

5 points per question

(1) 1 gallon   1 quart

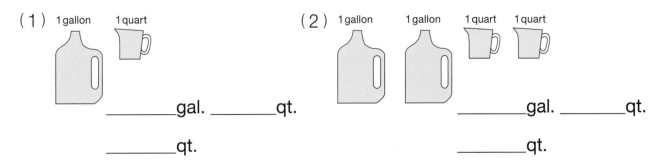

_____gal. _____qt.

_____qt.

(2) 1 gallon   1 gallon   1 quart   1 quart

_____gal. _____qt.

_____qt.

Reading
DAY
45

Review
Youth Olympic Games

Level ☆☆☆

Date
/   /

Name

① Read the passage. Then answer the questions using words from the 100 points for comple passage.

> The Olympic Games are a sporting event that started in ancient Greece. But it wasn't until 2010 that the Youth Olympic Games was launched. The Youth Olympic Games began at the Singapore Olympics. Its goal is to bring together the world's best young athletes to play and compete in sports, as well as to learn from new cultures. Even non-athletes participate as volunteers, reporters, torch bearers, and more.
>
> For the sports competitions, there are three age groups that can compete: fifteen- and sixteen-year-olds, sixteen- and seventeen-year-olds, and seventeen- and eighteen-year-olds. In the summer, kids play twenty-eight sports—sometimes with boys and girls on the same team. Talented kids from all around the world come together every Olympics, which take place every four years and last for ten to twelve days. In 2010, there were 3,600 athletes. To make sure that kids from every part of the world can play, there are four spots saved for each nation. Some of the sports include pole vaulting, gymnastics, diving, and tennis.

(1) What are the Olympic Games?

The Olympic Games are a _____ _____ that started in _____ Greece.

(2) Who can compete at the Youth Olympic Games?

There are _____ age groups that can compete: _____

year-olds, _____ year-olds, and _____

_____ year-olds.

(3) Where were the first Youth Olympic Games?

The first Youth Olympic Games were in _____.

(4) When were the first Youth Olympic Games?

The first Youth Olympic Games were in _____.

(5) Why are four spots saved for each nation?

Four spots are saved for each nation to make sure that _____

_____.

(6) How long does the Youth Olympic Games last?

The Youth Olympic Games last for _____ to _____ days.

DAY 1, pages 1 & 2

1 Add.

(1) 100 + 55 = 155
(2) 107 + 20 = 127
(3) 130 + 36 = 166
(4) 125 + 43 = 168
(5) 135 + 45 = 180
(6) 120 + 6 = 126
(7) 117 + 6 = 123
(8) 127 + 9 = 136
(9) 137 + 4 = 141
(10) 126 + 56 = 182

2 Add.

(1) 100 + 100 = 200
(2) 200 + 200 = 400
(3) 300 + 300 = 600
(4) 236 + 123 = 359
(5) 236 + 152 = 388
(6) 236 + 129 = 365
(7) 324 + 427 = 751
(8) 324 + 429 = 753
(9) 135 + 317 = 452
(10) 408 + 135 = 543

① Read each sentence aloud. Then trace the preposition.

(1) We walk to the park.
(2) The monkey swings on the branch.
(3) She met me at the pool.
(4) The cat hid under the couch.
(5) The dog jumps in the lake.
(6) He ran from the bookstore.
(7) We wait for the bus.
(8) They eat the box of candy.
(9) The birds perch near the fountain.
(10) We sit beside our camp counselor.

② Circle the preposition in each sentence below.

(1) I jogged (beside) my brother.
(2) We danced (to) the music.
(3) She jumped (on) the diving board.
(4) The man asked (for) three hot dogs.
(5) The turtle swam (under) the bridge.
(6) (In) the rollerskating rink, the skater spun around.
(7) He sunk a basket (from) mid-court.
(8) (At) the field, the team captains chose teammates.
(9) He wrote a book (of) short stories.
(10) (Near) the finish line, people cheered.

DAY 2, pages 3 & 4

1 Add.

(1) 162 + 154 = 316
(2) 251 + 190 = 441
(3) 136 + 272 = 408
(4) 580 + 150 = 730
(5) 360 + 382 = 742
(6) 250 + 371 = 621
(7) 275 + 305 = 580
(8) 275 + 16 = 291
(9) 275 + 26 = 301
(10) 275 + 36 = 311

2 Add.

(1) 229 + 76 = 305
(2) 317 + 85 = 402
(3) 264 + 67 = 331
(4) 293 + 394 = 687
(5) 298 + 394 = 692
(6) 475 + 396 = 871
(7) 375 + 487 = 862
(8) 367 + 262 = 629
(9) 178 + 239 = 417
(10) 141 + 259 = 400

① Trace the words in the "How to Read a Dictionary" passage below and read the definitions.

Syllable — a unit of sound that forms a part of a whole word.

Definition — the meaning of a word. If a word has more than one meaning, each meaning is listed with numbers.

Part of Speech — the type of word, such as a noun, verb, or adjective.

Pronunciation — how a word sounds.

gov-ern (guhv-urn) *verb*
To control a country or organization using laws.

gov-ern-ment (guhv-urn-muhnt) *noun*
1. The control of a country, state or organization.
2. The people who govern.

gown (goun) *noun*
1. A woman's dress, as in *a ball gown.*
2. A loose robe worn by judges, graduating students and doctors.

② Choose words from the box below to complete the descriptions. Hint: You can use a word more than once.

pronunciation gown syllable government

(1) gown a woman's dress
(2) syllable a unit of sound
(3) pronunciation how a word sounds
(4) government the people who govern
(5) gown a loose robe

DAY 3, pages 5 & 6

1 Subtract.

(1) 123 − 62 = 61
(2) 150 − 72 = 78
(3) 138 − 53 = 85
(4) 140 − 10 = 130
(5) 160 − 40 = 120
(6) 180 − 30 = 150
(7) 154 − 32 = 122
(8) 156 − 56 = 100
(9) 156 − 48 = 108
(10) 174 − 18 = 156

2 Subtract.

(1) 145 − 38 = 107
(2) 245 − 38 = 207
(3) 146 − 52 = 94
(4) 342 − 38 = 304
(5) 467 − 87 = 380
(6) 467 − 59 = 408
(7) 600 − 200 = 400
(8) 740 − 300 = 440
(9) 740 − 320 = 420
(10) 744 − 410 = 334

① Trace the prefixes below. Then write the new words according to the example.

(1) non + slip = nonslip
(2) dis + like = dislike
(3) anti + toxic = antitoxic
(4) sub + title = subtitle
(5) mis + step = misstep

② Make words to match the definitions below by connecting two puzzle pieces.

non dis title toxic like anti mis sub slip step

(1) nonslip created to reduce or prevent slipping
(2) misstep a wrong step
(3) antitoxic made to prevent, reduce or stop toxins
(4) subtitle a secondary title that explains more about the story
(5) dislike not like something; feel disgust for something

DAY 4, pages 7 & 8

1 Subtract.

(1) 435 − 313 = 122
(2) 435 − 225 = 210
(3) 435 − 400 = 35
(4) 556 − 352 = 204
(5) 556 − 443 = 113
(6) 556 − 328 = 228
(7) 444 − 162 = 282
(8) 444 − 126 = 318
(9) 444 − 141 = 303
(10) 636 − 572 = 64

2 Subtract.

(1) 132 − 15 = 117
(2) 132 − 27 = 105
(3) 132 − 38 = 94
(4) 225 − 48 = 177
(5) 225 − 38 = 187
(6) 225 − 158 = 67
(7) 100 − 2 = 98
(8) 200 − 2 = 198
(9) 210 − 5 = 205
(10) 101 − 2 = 99

① An adjective describes a noun. An adverb describes a verb. Complete the table below according to the example.

adjective	adverb
bright	brightly
recent	recently
loud	loudly
clever	cleverly
wise	wisely
brave	bravely
gentle	gently

② Complete the sentences with a pair of words from the box below.

gentle/gently recent/recently wise/wisely clever/cleverly loud/loudly bright/brightly brave/bravely

(1) My brother is wise. He wisely eats many vegetables.
(2) The band is loud. The guitarist plays his solo loudly.
(3) In recent days, the baby grew new teeth. She recently started eating solid food.
(4) The mother gently licked her kitten. The kitten purred because of the gentle strokes.
(5) The moon was bright. It brightly lit the field.
(6) What a clever student! He cleverly solved the math problem.
(7) The brave firefighter fought the flames. She bravely saved the family's home.

DAY 5, pages 9 & 10

1 Fill in the missing multiples of 2 in the boxes below.

2 - 4 - 6 - 8 - 10 - 12 - 14 - 16 - 18

2 Say the number sentence aloud as you multiply by 2.

(1) 2 × 1 = 2 Two times one equals two
(2) 2 × 2 = 4 Two times two equals four
(3) 2 × 3 = 6 Two times three equals six
(4) 2 × 4 = 8 Two times four equals eight
(5) 2 × 5 = 10 Two times five equals ten
(6) 2 × 6 = 12 Two times six equals twelve
(7) 2 × 7 = 14 Two times seven equals fourteen
(8) 2 × 8 = 16 Two times eight equals sixteen
(9) 2 × 9 = 18 Two times nine equals eighteen

3 Fill in the missing multiples of 3 in the boxes below.

3 - 6 - 9 - 12 - 15 - 18 - 21 - 24 - 27

4 Say the number sentence aloud as you multiply by 3.

(1) 3 × 1 = 3 Three times one equals three
(2) 3 × 2 = 6 Three times two equals six
(3) 3 × 3 = 9 Three times three equals nine
(4) 3 × 4 = 12 Three times four equals twelve
(5) 3 × 5 = 15 Three times five equals fifteen
(6) 3 × 6 = 18 Three times six equals eighteen
(7) 3 × 7 = 21 Three times seven equals twenty-one
(8) 3 × 8 = 24 Three times eight equals twenty-four
(9) 3 × 9 = 27 Three times nine equals twenty-seven

① Read the passage. Then match the words in bold with the words with similar meaning below.

It was a **dismal** day and Gary was bored. He thought it would be **amusing** to scare his little sister. He hid in her closet, and when she came in, he jumped out and yelled "Boo!" She screamed and fell back and scraped her arm. Gary didn't mean to **harm** her, but she began to cry. He got a bandage to help **mend** the scrape. He apologized and she **responded**, "I forgive you but watch out—I'm going to get even!" Gary now lived in fear! He surely wasn't bored anymore.

(1) reply (responded)
(2) fix (mend)
(3) funny (amusing)
(4) gloomy (dismal)
(5) hurt (harm)

② Read the sentence then put a (✓) next to the phrase that is similar.

(1) interested in books
(✓) likes to read ( ) thinks about movies
(2) shares her treat
(✓) offers her popcorn ( ) eats her own lunch
(3) breaks it in two
(✓) snaps it in half ( ) smashes it to pieces
(4) slips on the grass
( ) sits on the grass (✓) slides on the grass
(5) dark hair and eyes
(✓) black hair and brown eyes ( ) black hair and blue eyes

DAY 6, pages 11 & 12

1 Fill in the missing multiples of 4 in the boxes below.

4 - 8 - 12 - 16 - 20 - 24 - 28 - 32 - 36

2 Say the number sentence aloud as you multiply by 4.

(1) 4 × 1 = 4 Four times one equals four
(2) 4 × 2 = 8 Four times two equals eight
(3) 4 × 3 = 12 Four times three equals twelve
(4) 4 × 4 = 16 Four times four equals sixteen
(5) 4 × 5 = 20 Four times five equals twenty
(6) 4 × 6 = 24 Four times six equals twenty-four
(7) 4 × 7 = 28 Four times seven equals twenty-eight
(8) 4 × 8 = 32 Four times eight equals thirty-two
(9) 4 × 9 = 36 Four times nine equals thirty-six

3 Fill in the missing multiples of 5 in the boxes below.

5 - 10 - 15 - 20 - 25 - 30 - 35 - 40 - 45

4 Say the number sentence aloud as you multiply by 5.

(1) 5 × 1 = 5 Five times one equals five
(2) 5 × 2 = 10 Five times two equals ten
(3) 5 × 3 = 15 Five times three equals fifteen
(4) 5 × 4 = 20 Five times four equals twenty
(5) 5 × 5 = 25 Five times five equals twenty-five
(6) 5 × 6 = 30 Five times six equals thirty
(7) 5 × 7 = 35 Five times seven equals thirty-five
(8) 5 × 8 = 40 Five times eight equals forty
(9) 5 × 9 = 45 Five times nine equals forty-five

① Read the passage. Then match the words in bold with the words with opposite meanings below.

My family was planning a tubing trip down the Delaware River. My dad wanted to **purchase** an **enormous** raft for himself so he could stretch out. He searched and searched for the **perfect** raft. He found a huge raft that came with a **powerful** pump. Even so, when we went to the river, the raft took forever to blow up. When we finally got into the water, our tubes floated quickly, but my dad's heavy raft got stuck on every rock! As we **vanished** down the river, we yelled, "You forgot to buy paddles!"

(1) weak (powerful)
(2) small (enormous)
(3) appeared (vanished)
(4) sell (purchase)
(5) faulty (perfect)

② Read the sentence then put a check (✓) next to the contrasting phrase.

(1) likes to sing
( ) enjoys being in a chorus (✓) loves to play piano
(2) bakes some bread
(✓) eats some bread ( ) puts bread in oven
(3) glues the pieces together
(✓) breaks it apart ( ) pastes the parts
(4) bikes fast
( ) speeds away (✓) stands to watch
(5) bushy tail wags
( ) hairy tail thumps (✓) dog sleeps soundly

DAY 7, pages 13 & 14

1 Fill in the missing multiples of 6 in the boxes below.

6 - 12 - 18 - 24 - 30 - 36 - 42 - 48 - 54

2 Say the number sentence aloud as you multiply by 6.

(1) 6 × 1 = 6 Six times one equals six
(2) 6 × 2 = 12 Six times two equals twelve
(3) 6 × 3 = 18 Six times three equals eighteen
(4) 6 × 4 = 24 Six times four equals twenty-four
(5) 6 × 5 = 30 Six times five equals thirty
(6) 6 × 6 = 36 Six times six equals thirty-six
(7) 6 × 7 = 42 Six times seven equals forty-two
(8) 6 × 8 = 48 Six times eight equals forty-eight
(9) 6 × 9 = 54 Six times nine equals fifty-four

3 Fill in the missing multiples of 7 in the boxes below.

7 - 14 - 21 - 28 - 35 - 42 - 49 - 56 - 63

4 Say the number sentence aloud as you multiply by 7.

(1) 7 × 1 = 7 Seven times one equals seven
(2) 7 × 2 = 14 Seven times two equals fourteen
(3) 7 × 3 = 21 Seven times three equals twenty-one
(4) 7 × 4 = 28 Seven times four equals twenty-eight
(5) 7 × 5 = 35 Seven times five equals thirty-five
(6) 7 × 6 = 42 Seven times six equals forty-two
(7) 7 × 7 = 49 Seven times seven equals forty-nine
(8) 7 × 8 = 56 Seven times eight equals fifty-six
(9) 7 × 9 = 63 Seven times nine equals sixty-three

① Read the passage. Then choose the words in bold from the passages to complete the definitions below.

Long ago, some house mice met to **consider** what **measures** they could take to **outsmart** their **enemy**, the cat. Some said this, and some said that. At last a young mouse got up and said he had a **proposal** to make, that he thought would meet the **challenge**.

"You will all agree," said he, "that our **chief** danger is the **sly** way in which the enemy attacks us. Now, if we could receive some **signal** that she is coming, we could easily **escape** from her."

(1) proposal a suggestion
(2) sly clever, tricky, or sneaky
(3) signal a sound or motion made to give a warning or an order
(4) challenge a difficult task or a contest
(5) chief most important
(6) measures plans or courses of action
(7) consider to think carefully about something
(8) outsmart defeat by being clever
(9) enemy opponent or rival
(10) escape break free

DAY 8, pages 15 & 16

1 Fill in the missing multiples of 8 in the boxes below.

8 - 16 - 24 - 32 - 40 - 48 - 56 - 64 - 72

2 Say the number sentence aloud as you multiply by 8.

(1) 8 × 1 = 8 Eight times one equals eight
(2) 8 × 2 = 16 Eight times two equals sixteen
(3) 8 × 3 = 24 Eight times three equals twenty-four
(4) 8 × 4 = 32 Eight times four equals thirty-two
(5) 8 × 5 = 40 Eight times five equals forty
(6) 8 × 6 = 48 Eight times six equals forty-eight
(7) 8 × 7 = 56 Eight times seven equals fifty-six
(8) 8 × 8 = 64 Eight times eight equals sixty-four
(9) 8 × 9 = 72 Eight times nine equals seventy-two

3 Fill in the missing multiples of 9 in the boxes below.

9 - 18 - 27 - 36 - 45 - 54 - 63 - 72 - 81

4 Say the number sentence aloud as you multiply by 9.

(1) 9 × 1 = 9 Nine times one equals nine
(2) 9 × 2 = 18 Nine times two equals eighteen
(3) 9 × 3 = 27 Nine times three equals twenty-seven
(4) 9 × 4 = 36 Nine times four equals thirty-six
(5) 9 × 5 = 45 Nine times five equals forty-five
(6) 9 × 6 = 54 Nine times six equals fifty-four
(7) 9 × 7 = 63 Nine times seven equals sixty-three
(8) 9 × 8 = 72 Nine times eight equals seventy-two
(9) 9 × 9 = 81 Nine times nine equals eighty-one

① Read the passage. Then choose words from the passage to complete the definitions below.

The young mouse went on to **explain** how they would know if the cat was nearby.

"A small bell should be **acquired**, and **attached** with a ribbon around the neck of the cat. Then we will always know when she is around, and can easily **retreat** while she is in the **neighborhood**."

All the mice **applauded** for the smart **scheme**, except for the old mouse. He got up and said, "That is all very well, but who will bell the cat?"

The mice looked at one another and **grumbled** their excuses: "I have a bad back!" "I'm afraid of bells!" "I'm busy later."

Even the young mouse said, "I came up with the idea! I shouldn't have to bell the cat!" Then all the mice fell silent and the old mouse **concluded**.

"It is easy to think of **impossible** solutions."

(1) neighborhood an area around a place, person or object
(2) concluded judged or arrived at a judgement
(3) grumbled complained, whined or protested
(4) attached joined or connected to something or someone
(5) retreat to back down or withdraw from an enemy
(6) acquired bought or gotten
(7) impossible not able to occur, exist or be done
(8) scheme a plan
(9) applauded clapped
(10) explain make an idea clear by describing it in detail

DAY 9, pages 17 & 18

1 Fill in the missing multiples of 1 in the boxes below.

1 - 2 - 3 - 4 - 5 - 6 - 7 - 8 - 9

2 Say the number sentence aloud as you multiply by 1.

(1) 1 × 1 = 1 One times one equals one
(2) 1 × 2 = 2 One times two equals two
(3) 1 × 3 = 3 One times three equals three
(4) 1 × 4 = 4 One times four equals four
(5) 1 × 5 = 5 One times five equals five
(6) 1 × 6 = 6 One times six equals six
(7) 1 × 7 = 7 One times seven equals seven
(8) 1 × 8 = 8 One times eight equals eight
(9) 1 × 9 = 9 One times nine equals nine

3 Fill in the missing multiples of 10 in the boxes below.

10 - 20 - 30 - 40 - 50 - 60 - 70 - 80 - 90

4 Say the number sentence aloud as you multiply by 10.

(1) 10 × 1 = 10 Ten times one equals ten
(2) 10 × 2 = 20 Ten times two equals twenty
(3) 10 × 3 = 30 Ten times three equals thirty
(4) 10 × 4 = 40 Ten times four equals forty
(5) 10 × 5 = 50 Ten times five equals fifty
(6) 10 × 6 = 60 Ten times six equals sixty
(7) 10 × 7 = 70 Ten times seven equals seventy
(8) 10 × 8 = 80 Ten times eight equals eighty
(9) 10 × 9 = 90 Ten times nine equals ninety

(1) Read the passage and the vocabulary words defined below. Complete the passage using the vocabulary words.

Once upon a time there was a poor but very good little girl who lived alone with her mother. They had nothing in the house to eat.

So the child went to (1) forage in the forest, and there she met an (2) elderly woman. The old woman saw she was (3) distress because the girl looked skinny and (4) frail. The old woman (5) presented her with a pot which had (6) magical powers. If one said to it, "Boil, little pot!" it would cook sweet soup; and when one (7) commanded, "Stop, little pot!" it would (8) immediately stop boiling. The little girl took the pot home to her mother, and now their (9) poverty was at an end, for they could have sweet broth (10) whenever they wanted.

distress: in extreme pain or sadness
forage: search widely for food, drink or needed items
frail: weak and delicate
immediately: at once; instantly
magical: using magic
poverty: the state of being very poor
presented: gave or offered
whenever: at any time
elderly: an old or aging person
commanded: gave an order

DAY 10, pages 19 & 20

1 Multiply.

(1) 2 × 2 = 4
(2) 2 × 4 = 8
(3) 2 × 6 = 12
(4) 2 × 8 = 16
(5) 2 × 9 = 18
(6) 3 × 5 = 15
(7) 3 × 6 = 18
(8) 3 × 7 = 21
(9) 3 × 2 = 6
(10) 3 × 8 = 24
(11) 4 × 3 = 12
(12) 4 × 5 = 20
(13) 4 × 7 = 28
(14) 4 × 9 = 36
(15) 4 × 2 = 8
(16) 5 × 2 = 10
(17) 5 × 4 = 20
(18) 5 × 5 = 25
(19) 5 × 7 = 35
(20) 5 × 9 = 45
(21) 6 × 4 = 24
(22) 6 × 5 = 30
(23) 6 × 6 = 36
(24) 6 × 8 = 48
(25) 6 × 3 = 18
(26) 7 × 3 = 21
(27) 7 × 5 = 35
(28) 7 × 6 = 42
(29) 7 × 8 = 56
(30) 7 × 9 = 63
(31) 8 × 3 = 24
(32) 8 × 5 = 40
(33) 8 × 7 = 56
(34) 8 × 9 = 72
(35) 8 × 3 = 24
(36) 9 × 5 = 45
(37) 9 × 6 = 54
(38) 9 × 7 = 63
(39) 9 × 4 = 36
(40) 9 × 9 = 81
(41) 1 × 2 = 2
(42) 1 × 4 = 4
(43) 1 × 6 = 6
(44) 1 × 8 = 8
(45) 1 × 9 = 9
(46) 10 × 3 = 30
(47) 10 × 5 = 50
(48) 10 × 7 = 70
(49) 10 × 8 = 80
(50) 10 × 9 = 90

(1) Read the passage and the vocabulary words defined below. Complete the passage using the vocabulary words.

One day, the little girl (1) departed to visit a friend in the next town, and while she was (2) absent, her mother said: "Boil, little pot!" So the pot began to cook, and she soon ate all the soup she wished; but when the poor woman wanted to stop the pot, she (3) realized that she did not know the words. The pot (4) continually boiled and boiled, and soon it was (5) overflowing. As it boiled, the kitchen started to flood, then the house, and the next house, and soon the whole street! It seemed (6) likely to fill up all the town. Although there was a great (7) necessity for food, nobody wanted soup in their streets and living rooms; but nobody knew how to stop the pot. At last, when only a very small (8) cottage in all of the village was left unfilled by soup, the girl returned. She (9) sloshed through the soup and said at once: "Stop, little pot!" Instantly the pot (10) ceased. For the next year, whoever wished to enter the village had to eat his way through the soup!

absent: away or not present
ceased: came to an end
cottage: a small simple house
likely: probable or expected
necessity: a required or needed thing
overflowing: flooding or flowing over; so full that the contents spills over the rim
realized: became aware; noticed
sloshed: moved through liquid with a splashing sound
departed: went away
continually: constantly; again and again

DAY 11, pages 21 & 22

1 Multiply.

(1) 3 × 4 = 12
(2) 2 × 9 = 18
(3) 2 × 1 = 2
(4) 3 × 7 = 21
(5) 2 × 8 = 16
(6) 3 × 3 = 9
(7) 3 × 8 = 24
(8) 2 × 5 = 10
(9) 3 × 2 = 6
(10) 3 × 6 = 18
(11) 4 × 6 = 24
(12) 5 × 1 = 5
(13) 5 × 3 = 15
(14) 4 × 9 = 36
(15) 5 × 2 = 10
(16) 4 × 5 = 20
(17) 4 × 6 = 24
(18) 5 × 7 = 35
(19) 4 × 4 = 16
(20) 4 × 8 = 32
(21) 7 × 4 = 28
(22) 6 × 10 = 60
(23) 6 × 1 = 6
(24) 7 × 7 = 49
(25) 6 × 2 = 12
(26) 7 × 3 = 21
(27) 7 × 8 = 56
(28) 6 × 5 = 30
(29) 7 × 2 = 14
(30) 7 × 6 = 42
(31) 9 × 6 = 54
(32) 8 × 9 = 72
(33) 8 × 1 = 8
(34) 9 × 7 = 63
(35) 8 × 3 = 24
(36) 9 × 3 = 27
(37) 9 × 8 = 72
(38) 8 × 5 = 40
(39) 9 × 2 = 18
(40) 9 × 6 = 54
(41) 10 × 6 = 60
(42) 1 × 1 = 1
(43) 1 × 3 = 3
(44) 10 × 9 = 90
(45) 1 × 2 = 2
(46) 10 × 5 = 50
(47) 10 × 4 = 40
(48) 1 × 7 = 7
(49) 10 × 2 = 20
(50) 10 × 8 = 80

(1) Read the passage from "How the Wind Fills the Sails" by Dora Bernside. Use words from the passage to answer the questions below.

"What makes the vessel move on the river?" asked little Anna one day of her brother Harry.

"Why," said Harry, "it's the wind, of course, that fills the sails, and that pushes the vessel on. Come out on the lawn, and I will show you how it is done."

So Anna, Harry, and Bravo, all ran out on the lawn. Bravo was a dog, but he was always curious to see what was going on.

When they were on the lawn, Harry took out his handkerchief and told Anna to hold it by two of the corners while he held the other two corners.

As soon as they had done this, the wind made it swell out, and look just like a sail.

"Now you see how the wind fills the sails," said Harry.

(1) Who wants to know what makes vessels move on the river?
Anna wants to know what makes vessels move on the river.
(2) What pushes the vessel on the river?
The wind pushes the vessel on the river.
(3) Where does Harry suggest they go to test his idea?
Harry suggests they go out on the lawn.
(4) When does Harry take out his handkerchief?
Harry takes out his handkerchief when they are on the lawn.
(5) Who is always curious to see what is going on?
Bravo is always curious to see what is going on.
(6) What makes the handkerchief swell out?
The wind makes the handkerchief swell out.
(7) Where do Anna and Harry hold the handkerchief?
Anna and Harry hold the handkerchief at the corners.
(8) When the wind makes the handkerchief swell out, what does it look like?
The handkerchief looks like a sail.

DAY 12, pages 23 & 24

1 Multiply.

(1) 4 × 2 = 8
(2) 6 × 5 = 30
(3) 8 × 1 = 8
(4) 2 × 6 = 12
(5) 3 × 4 = 12
(6) 7 × 7 = 49
(7) 5 × 1 = 5
(8) 9 × 3 = 27
(9) 3 × 8 = 24
(10) 5 × 3 = 15
(11) 2 × 1 = 2
(12) 1 × 6 = 6
(13) 7 × 8 = 56
(14) 9 × 2 = 18
(15) 2 × 4 = 8
(16) 8 × 8 = 64
(17) 6 × 9 = 54
(18) 4 × 7 = 28
(19) 8 × 5 = 40
(20) 3 × 3 = 9
(21) 7 × 5 = 35
(22) 6 × 8 = 48
(23) 4 × 4 = 16
(24) 10 × 8 = 80
(25) 2 × 5 = 10
(26) 8 × 6 = 48
(27) 3 × 7 = 21
(28) 4 × 9 = 36
(29) 9 × 8 = 72
(30) 6 × 2 = 12
(31) 7 × 3 = 21
(32) 5 × 5 = 25
(33) 2 × 2 = 4
(34) 8 × 4 = 32
(35) 9 × 5 = 45
(36) 6 × 7 = 42
(37) 2 × 8 = 16
(38) 3 × 2 = 6
(39) 4 × 6 = 24
(40) 7 × 9 = 63
(41) 9 × 1 = 9
(42) 2 × 3 = 6
(43) 8 × 7 = 56
(44) 5 × 2 = 10
(45) 1 × 4 = 4
(46) 10 × 5 = 50
(47) 4 × 3 = 12
(48) 3 × 9 = 27
(49) 7 × 2 = 14
(50) 9 × 9 = 81

(1) Read the passage below. Use words from the passage to answer the questions below.

"Yes, but how does it make the ship go?" asked Anna.

"Well, now let go of the handkerchief, and see what becomes of it," said Harry.

So they both let go of it; and off the wind bore it up among the bushes by the side of the house.

In order to explain the matter still further to his sister, Harry made a little flat boat out of a shingle and put a mast in it, and on the mast he put a paper sail.

Then they went down to the river and launched it. Much to Anna's delight, the wind bore it far out towards the middle of the stream.

Bravo swam out, took it in his mouth, and brought it back; and Anna was at last quite satisfied that she knew how it is that the wind makes the vessel go on the river.

(1) Who made a little flat boat?
Harry made a little flat boat.
(2) What did Harry make a little flat boat out of?
Harry made a little flat boat out of a shingle, a mast, and a paper sail.
(3) Where did the handkerchief fly to?
The handkerchief flew to the bushes by the side of the house.
(4) When did the handkerchief fly to the bushes?
The handkerchief flew to the bushes when Anna and Harry both let go of it.
(5) Who swam out and brought back the little flat boat?
Bravo swam out and brought back the little flat boat.
(6) Where do they go to launch the boat?
They go down to the river to launch the boat.
(7) When was Anna convinced that she knew how vessels moved?
Anna was convinced when the wind bore the little ship out towards the middle of the stream.

DAY 13, pages 25 & 26

1 Write the missing number in each box.

(1) 2 × 3 = 6
(2) 2 × 5 = 10
(3) 2 × 7 = 14
(4) 3 × 5 = 15
(5) 4 × 7 = 28
(6) 5 × 8 = 40
(7) 6 × 6 = 36
(8) 7 × 8 = 56
(9) 8 × 9 = 72
(10) 9 × 7 = 63
(11) 10 × 7 = 70
(12) 2 × 2 = 4
(13) 3 × 2 = 6
(14) 4 × 3 = 12
(15) 4 × 4 = 16
(16) 5 × 5 = 25
(17) 3 × 6 = 18
(18) 4 × 7 = 28
(19) 8 × 8 = 64
(20) 9 × 9 = 81

2 Look at the example. Then write the missing number in each box.

Example
Multiplication
2 × 3 = 6
Division
6 ÷ 2 = 3

(1) 2 × 4 = 8 8 ÷ 2 = 4
(2) 2 × 7 = 14 14 ÷ 2 = 7
(3) 3 × 5 = 15 15 ÷ 3 = 5
(4) 4 × 5 = 20 20 ÷ 4 = 5
(5) 5 × 6 = 30 30 ÷ 5 = 6
(6) 6 × 8 = 48 48 ÷ 6 = 8
(7) 7 × 4 = 28 28 ÷ 7 = 4
(8) 8 × 7 = 56 56 ÷ 8 = 7
(9) 9 × 8 = 72 72 ÷ 9 = 8
(10) 10 × 8 = 80 80 ÷ 10 = 8

(1) Read the passage. Use words from the passage to answer the questions below.

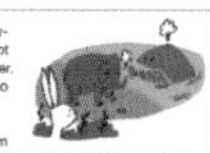

One day Rabbit was running along on the beach, hippety-hop, hippety-hop. He was going to a fine cabbage and carrot field. On the way he saw Whale and Elephant talking together. Rabbit said, "I'd like to know what they are talking about." So he crouched down behind some bushes and listened.

This is what Rabbit heard Whale say:

"You are the biggest thing on the land, Elephant, and I am the biggest thing in the sea. If we work together, we can scare all the animals with our size. Then we can rule all the animals in the world. We can have our own way about everything."

"Very good, very good," trumpeted Elephant. "That suits me. You keep the sea, and I will keep the land."

"That's a bargain," said Whale, as he swam away.

(1) Who wants to rule all the animals together?
Elephant and Whale want to rule all the animals.
(2) What does Elephant want to keep for himself?
Elephant wants to keep the land for himself.
(3) Where was Rabbit running?
Rabbit was running along on the beach.
(4) How does Rabbit hide?
Rabbit hides by crouching down behind some bushes.
(5) When did Rabbit see Whale and Elephant talking together?
Rabbit saw them talking together when he was going to the cabbage and carrot field.
(6) How do Elephant and Whale plan to rule all the animals?
Elephant and Whale plan to rule all the animals by scaring them with their size.
(7) Why does Whale want to rule all the animals?
Whale wants to rule all the animals so they can have their own way about everything.

DAY 14, pages 27 & 28

1 Look at the example. Then write the missing number in each box.

Example
Multiplication
2 × 3 = 6
Division
6 ÷ 3 = 2

(1) 2 × 2 = 4 4 ÷ 2 = 2
(2) 6 × 2 = 12 12 ÷ 2 = 6
(3) 3 × 3 = 9 9 ÷ 3 = 3
(4) 4 × 4 = 16 16 ÷ 4 = 4
(5) 5 × 5 = 25 25 ÷ 5 = 5
(6) 6 × 7 = 42 42 ÷ 6 = 7
(7) 7 × 5 = 35 35 ÷ 7 = 5
(8) 8 × 6 = 48 48 ÷ 8 = 6
(9) 9 × 7 = 63 63 ÷ 9 = 7
(10) 10 × 6 = 60 60 ÷ 10 = 6

2 Divide.

(1) 8 ÷ 2 = 4
(2) 10 ÷ 2 = 5
(3) 18 ÷ 2 = 9
(4) 6 ÷ 3 = 2
(5) 21 ÷ 3 = 7
(6) 8 ÷ 4 = 2
(7) 24 ÷ 4 = 6
(8) 10 ÷ 5 = 2
(9) 40 ÷ 5 = 8
(10) 12 ÷ 6 = 2
(11) 42 ÷ 6 = 7
(12) 49 ÷ 7 = 7
(13) 56 ÷ 8 = 7
(14) 54 ÷ 9 = 6
(15) 70 ÷ 10 = 7

(1) Read the passage. Use words from the passage to answer the questions below.

Rabbit laughed to himself. "Elephant and Whale may be big, but they won't rule me," he said.

Rabbit hopped off and soon came back with a very long and sturdy rope and his big drum. With one end of the rope, he walked up to Elephant who was munching on some tree leaves.

"Oh, dear Mr. Elephant," he said, "you are so big and strong; will you have the kindness to do me a favor?"

Elephant was flattered, and he trumpeted, "Certainly, certainly. What is it?"

"My cow is stuck in the mud on the shore, and I can't pull her out," said Rabbit. "You are so strong, I am sure you can get her out."

"Certainly, certainly," trumpeted Elephant.

"Thank you," said Rabbit. "Take this rope in your trunk, and I will tie the other end to my cow. Then I will beat my drum to let you know when to pull. You must pull as hard as you can, for the cow is very heavy."

"Huh!" trumpeted Elephant, "I'll pull her out by using my huge legs, or else I'll break the rope."

Rabbit tied the rope to Elephant's trunk and hopped away.

(1) Who says that Elephant is big and strong?
Rabbit says that Elephant is big and strong.
(2) What does Rabbit bring back with him?
Rabbit brings a very long and sturdy rope and his big drum.
(3) Where does Rabbit say he will tie the other end of the rope?
Rabbit says he will tie the other end to his cow.
(4) When is Elephant supposed to pull?
Elephant is supposed to pull when he hears Rabbit beat the drum.
(5) Why does Rabbit say he needs Elephant to pull his cow?
Rabbit says his cow is stuck in the mud on the shore, and he can't pull her out.
(6) How will Elephant pull out the cow?
Elephant will pull out the cow by using his huge legs.

DAY 15, pages 29 & 30

1 Divide.

(1) 8 ÷ 2 = 4
(2) 12 ÷ 2 = 6
(3) 10 ÷ 2 = 5
(4) 18 ÷ 3 = 6
(5) 24 ÷ 3 = 8
(6) 6 ÷ 3 = 2
(7) 20 ÷ 4 = 5
(8) 32 ÷ 4 = 8
(9) 12 ÷ 4 = 3
(10) 20 ÷ 5 = 4
(11) 35 ÷ 5 = 7
(12) 40 ÷ 5 = 8
(13) 30 ÷ 6 = 5
(14) 6 ÷ 6 = 1
(15) 42 ÷ 6 = 7
(16) 28 ÷ 7 = 4
(17) 14 ÷ 7 = 2
(18) 42 ÷ 7 = 6
(19) 32 ÷ 8 = 4
(20) 56 ÷ 8 = 7
(21) 64 ÷ 8 = 8
(22) 27 ÷ 9 = 3
(23) 54 ÷ 9 = 6
(24) 63 ÷ 9 = 7
(25) 50 ÷ 10 = 5

(1) Read the passage. Use words from the passage to answer the questions below.

Rabbit hopped till he came to the shore where Whale was. Making a bow, Rabbit said, "Mighty and wonderful Whale, will you do me a favor?"

"What is it?" asked Whale.

"My cow is stuck in the mud on the shore," said Rabbit, "and no one can pull her out. Of course you can do it. If you will be so kind as to help me, I will be very grateful."

"Certainly," said Whale, "certainly."

"Thank you," said Rabbit, "take hold of this rope in your teeth, and I will tie the other end to my cow. Then I will beat my big drum to let you know when to pull. You must pull as hard as you can, for my cow is very heavy."

"Never fear," said Whale, "I could pull a dozen cows out of the mud by flapping my huge tail."

"I am sure you could," said Rabbit politely. "Only be sure to begin gently. Then pull harder and harder till you get her out."

(1) Who does Rabbit call mighty and wonderful?
Rabbit calls Whale mighty and wonderful.
(2) What will Whale to hold in his teeth?
Whale will hold the rope in his teeth.
(3) Where does Rabbit say he will tie the other end of the rope?
Rabbit says he will tie the other end to his cow.
(4) When is Whale supposed to pull?
Whale is supposed to pull when he hears Rabbit beat the drum.
(5) Why does Rabbit say he needs Whale to pull his cow?
Rabbit says his cow is stuck in the mud on the shore, and he can't pull her out.
(6) How will Whale pull out the cow?
Whale will pull out the cow by flapping his huge tail.

DAY 16, pages 31 & 32

1 Divide.

(1) 12 ÷ 2 = 6
(2) 15 ÷ 3 = 5
(3) 28 ÷ 4 = 7
(4) 30 ÷ 5 = 6
(5) 24 ÷ 6 = 4
(6) 42 ÷ 7 = 6
(7) 56 ÷ 8 = 7
(8) 81 ÷ 9 = 9
(9) 8 ÷ 2 = 4
(10) 24 ÷ 3 = 8
(11) 45 ÷ 5 = 9
(12) 48 ÷ 6 = 8
(13) 21 ÷ 7 = 3
(14) 16 ÷ 4 = 4
(15) 35 ÷ 5 = 7
(16) 18 ÷ 6 = 3
(17) 63 ÷ 7 = 9
(18) 10 ÷ 2 = 5
(19) 27 ÷ 9 = 3
(20) 54 ÷ 6 = 9
(21) 72 ÷ 8 = 9
(22) 30 ÷ 10 = 3
(23) 54 ÷ 9 = 6
(24) 9 ÷ 3 = 3
(25) 32 ÷ 4 = 8

(1) Read the passage. Use words from the passage to answer the questions below.

Rabbit ran into the bushes and began to beat his drum. Then Whale began to pull, and Elephant began to pull. In a minute the rope tightened until it was stretched as hard as a bar of iron.

"This is a very heavy cow," said Elephant, "but I'll pull her out." Bracing his huge feet in the earth, he gave a giant pull.

Soon Whale found himself sliding toward the land.

"No cow in the mud is going to beat me," Whale said.

He got so mad at the cow that he went head first, down to the bottom of the sea. That was a pull! Elephant was jerked off his feet, and came slipping and sliding toward the sea. He was very angry. Kneeling down on the beach, he twisted the rope around his trunk. He trumpeted and gave his hardest pull. Whale popped up out of the water and they bumped into each other. Then, they finally saw that each had hold of the same rope!

Rabbit rolled out of the bushes laughing. "You two couldn't rule all the animals, because you can't rule me!"

(1) Who bumps into each other?
Elephant and Whale bump into each other.
(2) What does Elephant twist the rope around?
Elephant twists the rope around his trunk.
(3) Where does Elephant kneel down?
Elephant kneels down on the beach.
(4) When does Whale go down to the bottom of the sea?
Whale goes down to the bottom of the sea after he begins sliding toward the land.
(5) Why do Whale and Elephant bump into each other?
Whale and Elephant bump into each other because they each have hold of the same rope.
(6) How does Rabbit signal Whale and Elephant to start pulling?
Rabbit signals Whale and Elephant by beating his drum.

DAY 17, pages 33 & 34

1 Divide.

(1) 4 ÷ 2 = 2　(6) 6 ÷ 3 = 2　(11) 9 ÷ 4 = 2 R 1
(2) 5 ÷ 2 = 2 R 1　(7) 7 ÷ 3 = 2 R 1　(12) 17 ÷ 5 = 3 R 2
(3) 6 ÷ 2 = 3　(8) 8 ÷ 3 = 2 R 2　(13) 38 ÷ 6 = 6 R 2
(4) 7 ÷ 2 = 3 R 1　(9) 9 ÷ 3 = 3　(14) 25 ÷ 7 = 3 R 4
(5) 8 ÷ 2 = 4　(10) 10 ÷ 3 = 3 R 1　(15) 38 ÷ 8 = 4 R 6

2 Divide according to the example.

14 ÷ 3 = 4 R 2 —vertical form→ 3)14 = 4 R 2

3 Divide.

(1) 2)17 = 8 R 1　(4) 4)25 = 6 R 1　(7) 7)35 = 5
(2) 2)18 = 9　(5) 5)25 = 5　(8) 8)30 = 3 R 6
(3) 3)16 = 5 R 1　(6) 6)35 = 5 R 5　(9) 9)54 = 6

① Read the passage. Then answer the questions below.

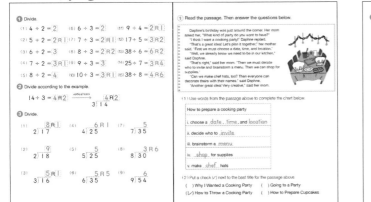

Daphne's birthday was just around the corner. Her mom asked her, "What kind of party do you want to have?"
"I think I want a cooking party!" Daphne replied.
"That's a great idea! Let's plan it together," her mother said. "First we must choose a date, time, and location."
"Well, we already know we need to be in our kitchen," said Daphne.
"That's right," said her mom. "Then we must decide who to invite and brainstorm a menu. Then we can shop for supplies."
"Can we make chef hats, too? Then everyone can decorate theirs with their names," said Daphne.
"Another great idea! Very creative," said her mom.

(1) Use words from the passage above to complete the chart below.

How to prepare a cooking party
i. choose a date, time, and location
ii. decide who to invite
iii. brainstorm a menu
iv. shop for supplies
v. make chef hats

(2) Put a check (✓) next to the best title for the passage above.

( ) Why I Wanted a Cooking Party　( ) Going to a Party
(✓) How to Throw a Cooking Party　( ) How to Prepare Cupcakes

DAY 18, pages 35 & 36

1 Read the word problem and write the number sentence below. Then answer the question.

(1) 353 people were in the amusement park. It started raining, so 95 people went home. How many people were left?

353 − 95 = 258　Ans. 258 people

(2) Mark's school has 186 boys and 189 girls. How many students go to Mark's school?

186 + 189 = 375　Ans. 375 students

(3) Mary read 157 pages yesterday. She read 256 pages today. How many pages did she read in all?

256 + 157 = 413　Ans. 413 pages

(4) A restaurant had 355 onions this morning. The cooks used 287 of them. How many onions does the restaurant have left?

355 − 287 = 68　Ans. 68 onions

(5) There are 187 boys and 178 girls at the park. Are there more boys or girls? How many more?

187 − 178 = 9　Ans. There are 9 more boys.

① Read the passage. Then answer the questions below.

Daphne thought about all her favorite treats: spaghetti and meatballs, watermelon, cookies and more. She decided she wanted to make homemade veggie pizza and ice cream sandwiches on her birthday. At the supermarket Daphne and her mom first got flour, salt and yeast for the pizza dough. Then they found the tomato sauce and shredded mozzarella cheese for the base. Then they picked out toppings: mushrooms, peppers, olives, eggplant, and onions. Last, they got ginger snap cookies and strawberry ice cream to make the ice cream sandwiches. Yum!

They also stopped at the arts and crafts store to get supplies for the chef hats. They bought white poster board, white tissue paper, tape, paper clips, and markers.

(1) Complete the chart below.

Pizza layer	Ingredients
Dough	(1) flour (2) salt (3) yeast
Base	(4) tomato sauce (5) shredded mozzarella cheese
(6) Topping	(7) mushrooms (8) peppers (9) olives (10) eggplant (11) onions

(2) Put a check (✓) next to the best title for the passage above.

( ) Ingredients for Pancakes　( ) Shopping for the Party
( ) What Not to Buy for Pizza　(✓) How to Cook Pizza

DAY 19, pages 37 & 38

1 Read the word problem and write the number sentence below. Then answer the question.

(1) Lily put 2 apples on each dish. There are 5 dishes on the table. How many apples are there in all?

Number of apples on a dish × Number of dishes = Total apples

2 × 5 = 10　Ans. 10 apples

(2) There are 3 vases in a room. Each vase has 3 flowers. How many flowers are there in all?

3 × 3 = 9　Ans. 9 flowers

(3) The teacher wants to give every student 2 pencils each. There are 9 students in the class. How many pencils will the teacher need?

2 × 9 = 18　Ans. 18 pencils

(4) In Jamal's homework group, there are 6 teams. Each team has 3 students. How many students are in Jamal's homework group?

3 × 6 = 18　Ans. 18 students

(5) Look at the picture below. How far is it from the red flag on the left to the blue flag on the right?

3 × 7 = 21　Ans. 21 feet

① Read the passage. Then answer the questions below.

First, make the dough: in a small bowl, dissolve yeast in warm water and add a dash of sugar. Meanwhile, in a mixing bowl add flour, salt, oil and spices. When yeast is bubbly, it's ready. Pour the yeast into the flour mixture and blend. Form your dough into a ball and place in a bowl that's been coated in oil. Let the dough rest 30 to 60 minutes. Roll out or stretch pizza dough into a pizza pan.

Then, add the base: spread sauce over stretched dough and sprinkle with cheese.

Last, put on toppings: choose your favorite vegetables as toppings and scatter them on the base.

Finally, ask a grown-up to bake your pizza at 450 degrees for 10 to 15 minutes.

(1) Complete the chart below.

Steps	Directions
Make the dough	(1) dissolve yeast in warm water and add sugar (2) mix together flour, salt, oil and spices (3) pour the bubbly yeast into the flour mixture (4) form the dough into a ball and let it rest (5) roll out the dough into a pizza pan
Add the base	(6) spread sauce over the stretched dough (7) sprinkle with cheese
(8) Put on toppings	(9) choose your favorite vegetables (10) scatter the toppings on the base

(2) Put a check (✓) next to the best title for the passage above.

( ) Pizza Disaster　( ) I Love Pizza
(✓) Veggie Pizza Recipe　( ) Hot Ovens

DAY 20, pages 39 & 40

1 Read the word problem and write the number sentence below. Then answer the question.

(1) There are 8 benches in the train station. 4 people can sit on each bench. How many people can sit on the benches in all?

4 × 8 = 32　Ans. 32 people

(2) The art teacher gives 5 sheets of colored paper to each student. There are 7 students in art class today. How many sheets of colored paper will the art teacher need?

5 × 7 = 35　Ans. 35 sheets

(3) We had 6 pieces of tape that were 4 inches long each. Just for fun, we connected them end to end. How long was our new piece of tape?

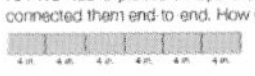

4 × 6 = 24　Ans. 24 inches

(4) The delivery man was making his rounds. He had 5 bags that each hold 6 packages. How many packages did he have?

5 × 6 = 30　Ans. 30 packages

(5) The width of a pool is 5 meters. The length is four times the width. How long is the pool?

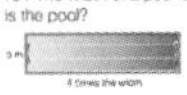

5 × 4 = 20　Ans. 20 m

① Read the passage. Then answer the questions below.

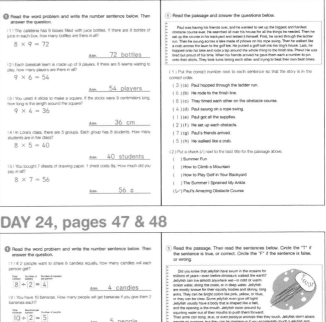

Mountains are like wrinkles on the face of the earth. Over time, mountains rise up, collide, crack and fold. Most scientists say that a mountain is land that rises 1,000 feet (300 meters) or more above the surrounding area.

The highest point on Earth is a folded mountain named Mount Everest. Mount Everest is in Nepal and is 29,035 feet (8,850 meters) high. The tallest mountain that is measured from top to bottom is Mauna Kea, which is in Hawaii. From the bottom, it is 33,474 feet (10,203 meters) tall but the mountain starts below the sea level, so it doesn't reach nearly as high above the earth's surface as Mount Everest. Mauna Kea is also a volcanic mountain, which forms when liquid rock from deep inside the Earth comes up through the ground and piles up. There are five main types of mountains: volcanic, dome, folded, plateau, and fault-block.

(1) Complete the chart below for Mount Everest.

Mountain name	Height	Location	Type of Mountain
Mount Everest	29,035 feet 8,850 meters	Nepal	folded

(2) Complete the chart below for Mauna Kea.

Mountain name	Height	Location	Type of Mountain
Mauna Kea	33,474 feet 10,203 meters	Hawaii	volcanic

(3) Put a check (✓) next to the best title for the passage above.

( ) How to Climb a Mountain　( ) Mauna Kea and Other Volcanoes
(✓) Facts About Mountains　( ) Mountain Animals

DAY 21, pages 41 & 42

1 Read the word problem and write the number sentence below. Then answer the question.

(1) Kim is going home for the holidays. She bought 6 boxes of candy as presents. Each box contains 6 candies. How many candies did she buy?

6 × 6 = 36　Ans. 36 candies

(2) 1 week is equal to 7 days. How many days are in 4 weeks?

7 × 4 = 28　Ans. 28 days

(3) There are 5 children in your house. You want to give them each 7 pieces of fruit. How many pieces of fruit will you need?

7 × 5 = 35　Ans. 35 pieces of fruit

(4) The bookstore has 5 boxes which each hold 6 books. How many books does the bookstore have?

6 × 5 = 30　Ans. 30 books

(5) Steve was playing with bricks. He piled 8 bricks up, one on top of the other. If each brick was 8 centimeters thick, how tall was his pile?

8 × 8 = 64　Ans. 64 cm

① Read the title of the story. Then put the correct number under each picture so that the story is in the correct order.

Making Tie-dyed T-shirts

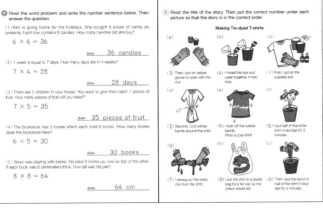

(a) (3) Then I put on rubber gloves to work with the dye.
(b) (4) I mixed the dye and water together in two tubs.
(c) (1) First, I got all the supplies out.
(d) (2) Second, I put rubber bands around the shirt.
(e) (9) I took off the rubber bands. What a cool shirt!
(f) (5) I put half of the white shirt in red dye for 5 minutes.
(g) (7) I wrung out the extra dye from the shirt.
(h) (8) I put the shirt in a plastic bag for a full day so the colors would set.
(i) (6) Then I put the second half of the shirt in blue dye for 5 minutes.

DAY 22, pages 43 & 44

1 Read the word problem and write the number sentence below. Then answer the question.

(1) The cafeteria has 9 boxes filled with juice bottles. If there are 8 bottles of juice in each box, how many bottles are there in all?

8 × 9 = 72　Ans. 72 bottles

(2) Each baseball team is made up of 9 players. If there are 6 teams waiting to play, how many players are there in all?

9 × 6 = 54　Ans. 54 players

(3) You used 4 sticks to make a square. If the sticks were 9 centimeters long, how long is the length around the square?

9 × 4 = 36　Ans. 36 cm

(4) In Lora's class, there are 5 groups. Each group has 8 students. How many students are in her class?

8 × 5 = 40　Ans. 40 students

(5) You bought 7 sheets of drawing paper. 1 sheet costs 8¢. How much did you pay in all?

8 × 7 = 56　Ans. 56 ¢

① Read the passage and answer the questions below.

Paul was having his friends over, and he wanted to set up the biggest and hardest obstacle course ever. He searched all over his house for all the things he needed. Then he set up the course in his backyard and tested it himself. First, he raced through the ladder run. Then he swung across a lake made of pillows on his rope swing. Then he walked like a crab across the lawn to the golf tee. He putted a golf ball into his dog's house. Last, he hopped onto his bike and rode a lap around the whole thing to the finish line. Phew! He was tired but proud of his time. When his friends arrived he gave them each a number to pin onto their shirts. They took turns timing each other and trying to beat their own best times.

(1) Put the correct number next to each sentence so that the story is in the correct order.

(3) (a) Paul hopped through the ladder run.
(6) (b) He rode to the finish line.
(8) (c) They timed each other on the obstacle course.
(4) (d) Paul swung on a rope swing.
(1) (e) Paul got all the supplies.
(2) (f) He set up each obstacle.
(7) (g) Paul's friends arrived.
(5) (h) He walked like a crab.

(2) Put a check (✓) next to the best title for the passage above.

( ) Summer Fun
( ) How to Climb a Mountain
( ) How to Play Golf in Your Backyard
( ) The Summer I Sprained My Ankle
(✓) Paul's Amazing Obstacle Course

DAY 23, pages 45 & 46

1 Read the word problem and write the number sentence below. Then answer the question.

(1) There are 9 bags with 5 apples in each. How many total apples are there?

5 × 9 = 45　Ans. 45 apples

(2) There are 7 bags with 8 stamps in each. How many total stamps are there?

8 × 7 = 56　Ans. 56 stamps

(3) Julia makes packets that have 4 candies each. If she gives 2 packets each to 7 people, how many candies is she giving away?

Number of candies per packet × Number of packets × Number of people = Total candies

4 × 2 × 7 = 56　Ans. 56 candies

(4) Jeff makes packets that have 3 candies each. If he gives 3 packets each to 5 people, how many candies is he giving away?

3 × 3 × 5 = 45　Ans. 45 candies

(5) The gardener gave each child 5 seeds to plant. If there are 48 children, how many seeds did the gardener give away?

5 × 48 = 240　Ans. 240 seeds

① Read the passage. Then read the sentences below. Circle the "T" if the sentence is true, or correct. Circle the "F" if the sentence is false, or wrong.

Amelia Earhart is one of the world's most well-known pilots. She was the first woman to fly alone over the Atlantic Ocean. When Amelia was growing up, she moved around a lot with her family. After high school, she worked as an army nurse in Canada. When she was twenty-three years old, she began to learn how to fly although her family didn't want her to fly. Two years later she bought her first plane. In 1928, she became the first woman to fly across the Atlantic Ocean even though she was only a passenger. Amelia became more and more set on piloting a plane across the Atlantic alone, which she was able to do four years later. She finished the trip in record time—fourteen hours and fifty-six minutes. In 1937, Amelia tried to fly around the world. After finishing most of her trip, she and her plane vanished and were never found.

(1) Amelia's family wanted her to learn to fly. T (F)
(2) When Amelia was a girl, she moved around a lot. (T) F
(3) After college, she worked as an army nurse. T (F)
(4) Amelia began to learn to fly when she was twenty-three years old. (T) F
(5) Amelia was the second woman to fly alone over the Atlantic Ocean. T (F)
(6) Amelia bought her first plane when she was twenty-five years old. (T) F
(7) In 1928, Amelia became the first female pilot. T (F)
(8) Amelia set a record when she flew across the Atlantic. (T) F
(9) Amelia tried to fly around the world. (T) F
(10) Amelia's plane vanished while she tried to fly around the world. (T) F

DAY 24, pages 47 & 48

1 Read the word problem and write the number sentence below. Then answer the question.

(1) If 2 people want to share 8 candies equally, how many candies will each person get?

Total candies ÷ Number of people = Number of candies per person

8 ÷ 2 = 4　Ans. 4 candies

(2) You have 10 bananas. How many people will get bananas if you give them 2 bananas each?

Total bananas ÷ Bananas per person = Number of people

10 ÷ 2 = 5　Ans. 5 people

(3) We have 21 pencils for my group today. If the 3 of us share them equally, how many pencils will each of us get?

21 ÷ 3 = 7　Ans. 7 pencils

(4) Your group in art class has 24 sheets of paper. If you want to give each person 3 sheets, how many people will get paper?

24 ÷ 3 = 8　Ans. 8 people

(5) If you divide 45 centimeters of ribbon into 9 equal parts, how long would each of those parts be?

45 ÷ 9 = 5　Ans. 5 cm

① Read the passage. Then read the sentences below. Circle the "T" if the sentence is true, or correct. Circle the "F" if the sentence is false, or wrong.

Did you know that jellyfish have swum in the oceans for millions of years—even before dinosaurs walked the earth? Jellyfish can live almost anywhere wet—in cold or warm ocean water, along the coast, or in deep water. Jellyfish are mostly known for their squishy bodies and skinny, long arms. They can be bright colors like pink, yellow, or blue, or they can be clear. Some jellyfish even give off light! Jellyfish usually have a body that is shaped like a bell, and the opening is the mouth. Jellyfish swim around by squirting water out of their mouths to push them forward. Their arms can sting, stun, or even paralyze animals that they touch. Jellyfish don't attack people on purpose, but they can be dangerous if you accidentally touch a jellyfish arm. Some arms can reach as long as 100 feet (33 meters)! Watch out!

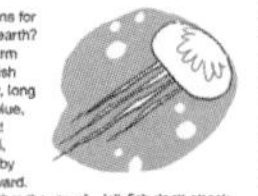

(1) Jellyfish are older than dinosaurs. (T) F
(2) Jellyfish can only live in warm ocean water. T (F)
(3) Most jellyfish have hard bodies and skinny, long arms. T (F)
(4) Jellyfish come in many different colors. (T) F
(5) All jellyfish give off light. T (F)
(6) Jellyfish don't attack people. (T) F
(7) Jellyfish have very short arms. T (F)
(8) Jellyfish move themselves forward by spitting water out of their mouths. (T) F
(9) Some jellyfish are shaped like a bassoon. T (F)
(10) Jellyfish can be dangerous. (T) F

DAY 25, pages 49 & 50

① Read the word problem and write the number sentence below. Then answer the question.

(1) There are 20 cookies and 6 children. If they divide the cookies equally and everyone get 3 cookies. How many cookies remain?

$20 \div 6 = 3 R 2$

Ans. **2** cookies remain

(2) There are 30 cookies. If 7 children get 4 cookies each, how many cookies remain?

$30 \div 7 = 4 R 2$

Ans. **2** cookies remain

(3) Your mother is making lunch. She divides 27 kiwis into 6 lunch bags evenly. How many kiwis are in each bag, and how many remain?

$27 \div 6 = 4 R 3$

Ans. **4** kiwis in each bag, **3** kiwis remain

(4) Your brother is making lunch bags. He has 35 strawberries and puts 8 into each lunch bag. How many bags will he make, and how many strawberries will remain?

$35 \div 8 = 4 R 3$

Ans. **4** bags, **3** strawberries remain

(5) June's mother has 50 roses, and she wants to put 8 in each vase. How many vases can she make? How many roses will she have left over?

$50 \div 8 = 6 R 2$

Ans. **6** vases, **2** roses left over

① Read the passage. Then answer the questions using words from passage.

A crow named Cassius picked up some beautiful feathers left on the ground by the peacocks. He thought he would look better than the other crows if he stuck them into his own tail, so he did. In fact, Cassius thought he was now too fine to mix with the other crows. So he strutted off to the peacocks, and thought he'd be welcomed as one of them.

The peacocks at once saw through his disguise. They disliked Cassius for being so vain. So they began to peck him, and soon he was stripped of all his borrowed feathers.

Feeling a little naked and embarrassed, Cassius went sadly home. He wanted to join his old crow friends as if nothing had happened. But they remembered how he had mocked them. They chased him away and would have nothing to do with him.

"If you had been happy," said one crow, "to remain as nature made you, instead of trying to be what you are not, you would not have been refused by the others or disliked by your equals."

(1) Why did Cassius stick the peacock feathers in his tail?
He thought he would **look** **better** than the other crows.

(2) What happened when Cassius went to be with the peacocks?
The peacocks saw through his **disguise**.

(3) Why did the peacocks begin to peck Cassius?
The peacocks began to peck Cassius because they disliked him for being so **vain**.

② Complete the chart with words from the passage above.

Cause	Result or Effect
Because the peacocks didn't accept Cassius.	Cassius wanted to **join** his old crow friends as if **nothing** had happened.
Because Cassius mocked his friends, the crows.	The crows **chased** him away and would have **nothing** to do with him.

DAY 26, pages 51 & 52

① Answer the following questions about the number '37,456,812.'

(1) Fill the words in each box below.

ten-millions place	millions place	hundred-thousands place	ten-thousands place	thousands place	hundreds place	tens place	ones place
3	7	4	5	6	8	1	2

(2) Reading from the thousands place to the left, we see the **ten-thousands** place, the **hundred-thousands** place, the **millions** place and the **ten-millions** place.

(3) The 7 in 37,456,812 is in the **millions** place.

(4) The 4 in 37,456,812 is in the **hundred-thousands** place.

(5) The 3 in 37,456,812 is in the **ten-millions** place.

② Write the correct number in each box.

(1) 60,000 is the number you get from adding **6** ten-thousands.

(2) 260,000 is the number you get from adding **26** ten-thousands.

(3) 260,000 is the number you get from adding **260** thousands.

(4) 2,600,000 is the number you get from adding **260** ten-thousands.

(5) 2,600,000 is the number you get from adding **2,600** thousands.

③ Compare the numbers below. Write < or > in the boxes.

(1) 5,000 **>** 3,000   (3) 101,101 **<** 101,110

(2) 690,000 **>** 67,000   (4) 1,801,012 **>** 1,800,901

① Match the cause with the effect.

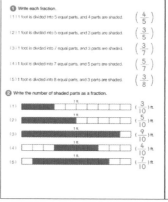

Cause		Effect
(1) Jessica was thirsty...		Ⓐ ...so instead she played guitar.
(2) The elephant got scared...		Ⓑ ...my mom got a broom and carefully swept it up.
(3) The sun was shining...		Ⓒ ...so he ran away.
(4) Mary hated drawing...		Ⓓ ...so they carried it back to their anthill.
(5) The beach was dirty...		Ⓔ ...he dug up his bone.
(6) The ants found a crumb...		Ⓕ ...she got a glass of juice.
(7) I broke a glass by accident...		Ⓖ ...so she took a nap.
(8) She was tired...		Ⓗ ...soon it would be dark outside.
(9) The dog was hungry...		Ⓘ ...so we couldn't have a picnic.
(10) It was raining...		Ⓙ ...so we got trash bags to clean it up.

DAY 27, pages 53 & 54

① How many parts of the whole figure are shaded?

(1) $\frac{1}{2}$ (3) $\frac{1}{4}$ (5) $\frac{1}{5}$ (7) $\frac{1}{5}$

(2) $\frac{1}{3}$ (4) $\frac{1}{4}$ (6) $\frac{1}{5}$ (8) $\frac{1}{5}$

② Each piece of tape is 1 foot long and has been divided into equal parts. How many parts of the whole piece of tape are shaded in each figure below?

(1) $\frac{1}{2}$ ft.
(2) $\frac{1}{3}$ ft.
(3) $\frac{1}{4}$ ft.
(4) $\frac{1}{6}$ ft.
(5) $\frac{3}{8}$ ft.
(6) $\frac{3}{5}$ ft.

① Read the passage. Then answer the questions below.

Every year, the Iowa State Fair celebrates butter in many special ways—from showing people how to make a butter cookie competition. But the greatest attraction is the famous Butter Cow.

The Butter Cow starts with a frame made out of wood, metal, wire and steel. The frame is put on a cooler that is about forty degrees. Then about six hundred pounds of pure Iowa butter is added layer upon layer. Once enough butter is on the frame, the artist carves a life-size cow that is about 5.5 feet (1.7 meters) high and 8 feet (2.4 meters) long.

John Karl Daniels made the first Iowa State Fair butter cow in 1911. Since then, there have only been four other people who have sculpted the Butter Cow. In 1960, Norma "Duffy" Lyon was the first woman to sculpt the Butter Cow, and she did it every year for forty-six years!

The Iowa State Fair Butter Cow turned one hundred years old in 2011. That year, the fair made one hundred cows from all kinds of objects like cars, flowers and sand, and placed these throughout the fairgrounds.

(1) What is the main idea of the second paragraph? Put a check (✓) next to the correct sentence below.
( ) The Iowa State Fair celebrates butter.
(✓) The Butter Cow is made by sculpting butter on a frame.
( ) The Butter Cow must be made in a 40 degree cooler.

(2) What is the main idea of the last paragraph? Put a check (✓) next to the correct sentence below.
( ) Every year the Iowa State Fair has the Butter Cow.
( ) Cows can be made from sand, too.
(✓) The Iowa State Fair celebrated 100 years of the Butter Cow in 2011.

(3) Put a check (✓) next to the best title for the passage below.
(✓) The Iowa State Fair's Butter Cow
( ) How to Make a Butter Cow
( ) New Uses for Butter

DAY 28, pages 55 & 56

① Write each fraction.

(1) 1 foot is divided into 5 equal parts, and 4 parts are shaded. $\left(\frac{4}{5}\right)$

(2) 1 foot is divided into 5 equal parts, and 3 parts are shaded. $\left(\frac{3}{5}\right)$

(3) 1 foot is divided into 7 equal parts, and 3 parts are shaded. $\left(\frac{3}{7}\right)$

(4) 1 foot is divided into 7 equal parts, and 5 parts are shaded. $\left(\frac{5}{7}\right)$

(5) 1 foot is divided into 8 equal parts, and 3 parts are shaded. $\left(\frac{3}{8}\right)$

② Write the number of shaded parts as a fraction.

(1) $\frac{3}{10}$ ft.
(2) $\frac{5}{10}$ ft.
(3) $\frac{9}{10}$ ft.
(4) $\frac{4}{10}$ ft.
(5) $\frac{7}{10}$ ft.

① Read the passage. Then answer the questions below.

Australia is a special place. It is the only country that is also a continent. A continent is a large piece of land on the globe, for example North America or Asia. Australia has many different kinds of land—from desert to rainforest. Australia also has many rare animals and plants that live there and nowhere else, like the kangaroo.

One of the most famous areas of Australia is the outback. It has the country's hottest weather and very few animals are able to live there. There are large deserts with very little water and almost no plants. Native people, called Aborigines, have learned to live in this difficult climate.

However, most people live on the edges of the country near the coast where the weather is more mild. Australians love to play sports in Australia's warm weather and great outdoors. Australians swim, surf, sail, and play soccer. Australians have even invented their own type of football.

(1) What is the main idea of the first paragraph? Put a check (✓) next to the correct sentence below.
( ) Australia has a large desert.
( ) The weather in Australia changes often.
(✓) Australia is a unique place.

(2) What is the main idea of the second paragraph? Put a check (✓) next to the correct sentence below.
( ) Australia has rare animals.
( ) Native people of Australia are called Aborigines.
(✓) The outback is a tough place.

(3) Put a check (✓) next to the best title for the passage above.
( ) How to Travel to Australia
(✓) A Guide to Australia
( ) Australian Sports

DAY 29, pages 57 & 58

① Convert the measurements below.

(1) 1 mi. = **5,280** ft.   (9) 5,280 ft. = **1** mi.

(2) 1 mi. 3 ft. = **5,283** ft.   (10) 5,285 ft. = **1** mi. **5** ft.

(3) 1 mi. 80 ft. = **5,360** ft.   (11) 6,000 ft. = **1** mi. **720** ft.

(4) 1 mi. 700 ft. = **5,980** ft.   (12) 8,000 ft. = **1** mi. **2,720** ft.

(5) 2 mi. = **10,660** ft.   (13) 10,560 ft. = **2** mi.

(6) 2 mi. 20 ft. = **10,580** ft.   (14) 11,000 ft. = **2** mi. **440** ft.

(7) 3 mi. 10 ft. = **15,850** ft.   (15) 14,000 ft. = **2** mi. **3,440** ft.

(8) 3 mi. 220 ft. = **16,060** ft.   (16) 16,800 ft. = **2** mi. **5,040** ft.

② Order the following lengths from longest to shortest with the numbers 1 through 4.

(1) 1 mi.   5,370 ft.   1 mi. 50 ft.   5,300 ft.
(4)   (1)   (2)   (3)

(2) 10,550 ft.   2 mi.   1 mi. 600 ft.   16,000 ft.
(3)   (1)   (4)   (2)

① Read the passage. Then answer the questions below.

A red robin, whose throat was dry from singing, saw a large pitcher in the distance. Happily, he flew to it, but found that it held only a little water, and even that was too near the bottom to be reached.

The robin bent down and stretched his neck, but he had no luck. Next he tried to tip the pitcher, thinking that he would at least be able to catch some of the water as it fell out. But he was not strong enough to move it at all. He even asked his friend the worm to try to push the pitcher with him, but the worm was really no help.

He walked round and round the edge of the pitcher and sang a tune to help himself think. Then he saw some pebbles lying nearby. He had an idea! He picked up the pebbles and dropped them one by one into the pitcher. After many pebbles, he managed at last to raise the water up to the very top, and took a good, long drink.

(1) Put a check (✓) next to the words that describe the robin.
( ) strange   ( ) sad   ( ) dirty
(✓) thirsty   (✓) strong   ( ) musical
(✓) clever   (✓) red   ( ) mean

(2) What did the robin want?
The robin wanted some **water**.

(3) Who tried to help the robin?
The **worm** tried to help the robin.

(4) What did the robin do to help himself think?
The robin **sung** a tune to help him think.

(5) Does the robin achieve his goal?
The robin **does** achieve his goal.

DAY 30, pages 59 & 60

① 1 centimeter (cm) is equal to 10 millimeters (mm). How far is each box from the left side of the ruler?

1 mm   1 cm 1 mm   2 cm   5 cm 5 mm   7 cm 3 mm   10 cm

(1) 5 cm
(2) 14 cm 9 mm
(3) 8 cm 7 mm
(4) 10 cm 4 mm

② Along each dashed line, draw a line that fits the measurement given below.

(1) 5 cm
(2) 14 cm 9 mm
(3) 8 cm 7 mm
(4) 10 cm 4 mm

③ 1,000 meters (m) are equal to 1 kilometer (km). Convert the measurements below.

(1) 1 km = **1,000** m   (6) 1,000 m = **1** km

(2) 1 km 7 m = **1,007** m   (7) 1,500 m = **1** km **500** m

(3) 1 km 50 m = **1,050** m   (8) 1,950 m = **1** km **950** m

(4) 1 km 600 m = **1,600** m   (9) 2,001 m = **2** km **1** m

(5) 2 km 50 m = **2,050** m   (10) 3,030 m = **3** km **30** m

④ Order the following lengths from longest to shortest with the numbers 1 through 4.

(1) 1 km   998 m   1 km 101 m   1,110 m
(3)   (4)   (2)   (1)

(2) 1,850 m   2 km   1 km 780 m   2,020 m
(3)   (2)   (4)   (1)

① Read the passage. Then answer the questions below.

One winter day the brown bear saw the fox who was slinking along with some fish he had stole.

"Hey! Stop a minute! Where did you get those?" demanded the bear.

"Oh, my. Well, I've been out fishing and caught them," the fox lied.

So the bear wanted to learn to fish, too, and asked the fox to tell him how he was to set about it.

"Oh, it is quite easy," answered the fox, "and simple to learn. You've only got to go upon the ice, and cut a hole and stick your tail down through it, and hold it there as long as you can. Don't pay attention if it smarts a little; that's when the fish bite. The longer you hold it there, the more fish you'll get, and then all at once, out with it! You must pull sideways, and give a strong pull, too."

Well, the bear did as the fox said, and though he felt very cold, and his tail smarted very much, he kept it in a long, long time. He wanted all the fish in the pond! But at last he had to wish frozen in the ice, though of course he did not know that. When he yanked it out, his tail snapped right off, and that's why the bear walks around with a stumpy tail to this day!

(1) Which character had stolen some fish?
The **fox** had stolen some fish.

(2) What did the bear want?
The bear wanted the **fish** in the **pond**.

(3) Put a check (✓) next to the words that describe the bear.
(✓) hungry   ( ) tricky   ( ) slimy
( ) thirsty   ( ) greedy   (✓) brown
( ) clever   (✓) tired   ( ) skinny

(4) Put a check (✓) next to the words that describe the fox.
(✓) sneaky   ( ) unhappy   ( ) talented
(✓) heavy   ( ) strong   (✓) mean
(✓) clever   ( ) hot   (✓) sly

DAY 31, pages 61 & 62

① The figures below are made of squares that have 1-inch sides. What is the area of each figure?

(1) (1 in.²)   (3) (2 in.²)   (5) (4 in.²)

(2) (2 in.²)   (4) (3 in.²)

② The figures below are made of squares that have 1-centimeter sides. What is the area of each figure?

(1) (1 cm²)   (3) (2 cm²)   (5) (4 cm²)

(2) (2 cm²)   (4) (3 cm²)

① Read the passage below. Then answer the questions.

Once upon a time there were three children named Wendy, John, and Michael, who lived with their father and mother in London. One evening the father and mother were readying a party, and the mother, after lighting the dim lamp in the nursery and kissing the children good-night, went away. Later that evening, a little boy climbed in through the window. His name was Peter Pan. He was a curious little fellow, very conceited, very forgetful, and yet very lovable. The most remarkable thing about him was that he never grew up. There came fitting in through the window with him his fairy, whose name was Tinker Bell.

Peter Pan woke all the children up, and after he had sprinkled fairy dust on their shoulders, he took them away to Neverland, where he lived with a family of lost boys.

(1) In what city does this story take place?
The story takes place in **London**.

(2) In which characters' bedroom does the story begin?
The story begins in **Wendy, John and Michael** 's bedroom.

(3) Does the story start during the day or during the night?
The story starts during the **night**.

(4) Where does Peter Pan take the children?
Peter Pan takes the children to **Neverland**.

(5) What is the setting of the first paragraph?
The setting of the first paragraph is in the evening, in the children's **bedroom** in the city of **London**.

DAY 32, pages 63 & 64

① Read each scale. Then write the weight below.

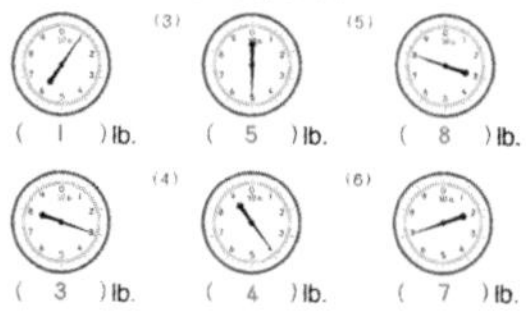

(1) (**1**) lb.   (**5**) lb.   (**8**) lb.

(2) (**3**) lb.   (**4**) lb.   (**7**) lb.

② 16 ounces (oz.) are equal to 1 pound (lb.). Convert the measurements below.

(1) 1 lb. = **16** oz.   (8) 16 oz. = **1** lb.

(2) 1 lb. 2 oz. = **18** oz.   (9) 20 oz. = **1** lb. **4** oz.

(3) 1 lb. 5 oz. = **21** oz.   (10) 23 oz. = **1** lb. **7** oz.

(4) 1 lb. 8 oz. = **24** oz.   (11) 22 oz. = **1** lb. **6** oz.

(5) 1 lb. 11 oz. = **27** oz.   (12) 25 oz. = **1** lb. **9** oz.

(6) 2 lb. 4 oz. = **36** oz.   (13) 35 oz. = **2** lb. **3** oz.

(7) 3 lb. 2 oz. = **50** oz.   (14) 49 oz. = **3** lb. **1** oz.

① Read the passage below. Then answer the questions.

Tinker Bell was jealous of the little girl Wendy, and she hurried ahead of Peter Pan and persuaded the lost boys that Wendy was a bird who might do them harm, and so one of the boys shot her with his bow and arrow.

When Peter Pan came and found Wendy lying lifeless upon the ground in the woods he was very angry, but he was also very quick-witted. So he told the boys that if they build a house around Wendy, she would be better. So they hurried to collect everything they had out of which they could make a house.

(1) What are the main events of this passage?
(a) Tinker Bell persuades the **lost boys** that Wendy might do them harm.
(b) One of the boys shoots Wendy with his **bow and arrow**.
(c) Peter Pan finds **Wendy** lying lifeless in the woods.
(d) Peter Pan **tells** the boys to **build** a house around Wendy.
(e) The boys **collect** everything to make a house.

(2) Why is Peter Pan angry?
Peter Pan is angry because one of the boys **shot** Wendy.

(3) Why do the boys build a house?
The boys build a house to make Wendy **better**.

② Read the passage below. Then answer the questions.

When the house was done, Peter Pan took John's hat for the chimney. The little house was so pleased to have such a capital chimney that smoke at once began to rise through the hat. All that night, Peter Pan walked up and down the front of Wendy's house to watch over her and keep her from danger while she slept.

What are the main events of this passage?
(a) Peter Pan takes John's **hat** to make the **chimney**.
(b) The house likes its chimney and puffs **smoke**.
(c) Peter **walks** up and down in front of Wendy's house.

DAY 33, pages 65 & 66

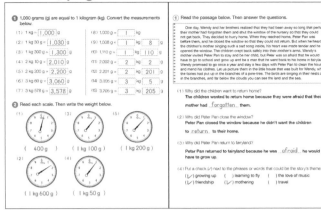

❶ 1,000 grams (g) are equal to 1 kilogram (kg). Convert the measurements below.

(1) 1 kg = 1,000 g
(2) 1 kg 30 g = 1,030 g
(3) 1 kg 300 g = 1,300 g
(4) 2 kg 10 g = 2,010 g
(5) 2 kg 200 g = 2,200 g
(6) 3 kg 60 g = 3,060 g
(7) 3 kg 578 g = 3,578 g
(8) 1,000 g = 1 kg
(9) 1,008 g = 1 kg 8 g
(10) 1,110 g = 1 kg 110 g
(11) 2,002 g = 2 kg 2 g
(12) 2,201 g = 2 kg 201 g
(14) 3,005 g = 3 kg 5 g
(15) 3,205 g = 3 kg 205 g

❷ Read each scale. Then write the weight below.

(1) (400 g)
(2) (1 kg 600 g)
(3) (1 kg 100 g)
(4) (1 kg 50 g)
(5) (1 kg 200 g)

① Read the passage below. Then answer the questions.

One day, Wendy and her brothers realized that they had been away so long that perhaps their mother had forgotten them and shut the window of the nursery so that they could not get back. They decided to hurry home. When they reached home, Peter Pan was before them, and he closed the window so that they could not return. But when he heard the children's mother singing such a sad song inside, his heart was made tender and he opened the window. The children crept back safely into their mother's arms. Wendy's mother invited Peter Pan to stay and be her child, but Peter was so afraid that he would have to go to school and grow up and be a man that he went back to his home in fairyland. Wendy promised to go once a year and stay a few days with Peter Pan to clean the house and mend his clothes. Let us picture them in the little house that was built for Wendy, which the fairies had put up in the branches of a pine-tree. The birds are singing in their nests and in the branches, and far below the clouds you can see the land and the sea.

(1) Why did the children want to return home?
The children wanted to return home because they were afraid that their mother had forgotten them.

(2) Why did Peter Pan close the window?
Peter Pan closed the window because he didn't want the children to return to their home.

(3) Why did Peter Pan return to fairyland?
Peter Pan returned to fairyland because he was afraid he would have to grow up.

(4) Put a check (✓) next to the phrases or words that could be the story's theme.
(✓) growing up ( ) learning to fly ( ) the love of music
(✓) friendship (✓) mothering ( ) travel

DAY 34, pages 67 & 68

❶ 1,000 milligrams (mg) are equal to 1 gram (g). Convert the measurements below.

(1) 1 g = 1,000 mg
(2) 3 g = 3,000 mg
(3) 7 g = 7,000 mg
(4) 1,000 mg = 1 g
(5) 4,000 mg = 4 g
(6) 5,000 mg = 5 g

❷ Order the following weights from heaviest to lightest with the numbers 1 through 4.

(1)	1 lb. (3)	17 oz. (2)	1 lb. 2 oz. (1)	15 oz. (4)
(2)	29 oz. (4)	2 lb. (3)	2 lb. 2 oz. (1)	33 oz. (2)
(3)	1 kg (3)	1,100 g (1)	1 kg 90 g (2)	900 g (4)
(4)	2 kg 300 g (1)	2,090 g (4)	2 kg 290 g (2)	2,110 g (3)
(5)	3 kg 850 g (3)	4 kg (1)	3,900 g (2)	3 kg 99 g (4)
(6)	1 g (3)	1,101 mg (1)	999 mg (4)	1,090 mg (2)
(7)	1 mg (4)	1 kg (1)	1 g (2)	100 mg (3)

① Read the sentences below. Circle the things that are being compared in the simile.

(1) The girl was as graceful as a swan.
(2) The boat drifted like a cloud.
(3) Jeff's cannonball into the pool was like a bomb.
(4) At the talent show, Robin was cool as a cucumber.
(5) When I looked down from the plane's window, the farmland looked like a quilt.
(6) My brother dances like a robot.
(7) The woman's red hair was like fire.
(8) My backpack feels as deep as the ocean.

② Choose words from the box below to finish the simile. Then circle the words that compare the two things.

oven molasses mountain fountain lightning

(1) The car drove as fast as lightning.
(2) The summer day was hot like an oven.
(3) The dog was drooling like a fountain.
(4) The football player was solid like a mountain.
(5) My chubby cat was as slow as molasses.

DAY 35, pages 69 & 70

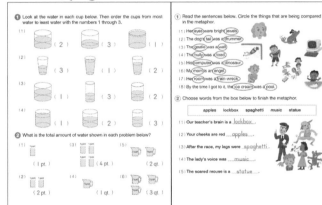

❶ Look at the water in each cup below. Then order the cups from most water to least water with the numbers 1 through 3.

(1) (2) (3) (1)
(2) (3) (1) (2)
(3) (1) (3) (2)
(4) (1) (2) (3)

❷ What is the total amount of water shown in each problem below?

(1) (1 pt.)
(2) (2 pt.)
(3) (4 pt.)
(4) (1 qt.)
(5) (2 qt.)
(6) (3 qt.)

① Read the sentences below. Circle the things that are being compared in the metaphor.

(1) Her eyes were bright jewels.
(2) The dog's tail was a drummer.
(3) The goalie was a wall.
(4) The bully was a toad.
(5) His computer was a dinosaur.
(6) My mom is an angel.
(7) Her room was a train-wreck.
(8) By the time I got to it, the ice cream was a pool.

② Choose words from the box below to finish the metaphor.

apples lockbox spaghetti music statue

(1) Our teacher's brain is a lockbox.
(2) Your cheeks are red apples.
(3) After the race, my legs were spaghetti.
(4) The lady's voice was music.
(5) The scared mouse is a statue.

DAY 36, pages 71 & 72

❶ Convert the measurements below.

(1) 1 qt. = 2 pt.
(2) 1 qt. 1 pt. = 3 pt.
(3) 2 qt. 1 pt. = 5 pt.
(4) 1 gal. = 4 qt.
(5) 1 gal. 3 qt. = 7 qt.
(6) 1 gal. = 8 pt.
(7) 1 gal. 2 qt. = 12 pt.
(8) 2 pt. = 1 qt.
(9) 4 pt. = 2 qt.
(10) 7 pt. = 3 qt. 1 pt.
(11) 5 qt. = 1 gal. 1 qt.
(12) 9 qt. = 2 gal. 1 qt.
(13) 8 pt. = 2 gal.
(14) 10 pt. = 1 gal. 1 qt.

❷ What is the total amount of water shown in each problem below?

(1) 1 gal. 2 qt.
6 qt.
(2) 3 qt.
6 pt.
(3) 2 gal. 1 qt.
9 qt.
(4) 1 qt. 1 pt.
3 pt.

① Read the poem below. Then answer the questions.

Tis a lesson you should heed,
Try, try, try again;
If at first you don't succeed,
Try, try, try again.
Once or twice though you should fail
Try again;
If you would at last prevail,
Try again.
If we strive, 'tis no disgrace
Though we may not win the race;
What should you do in that case?
Try again.
If you find your task is hard,
Try again;
Time will bring you your reward,
Try again.
All that other folks can do,
With your patience should not you?
Only keep this rule in view—
Try again.

(1) Put a check (✓) next to the phrases that describe the poem's main idea.
(✓) hard work will pay off (✓) do your best
( ) being lazy is fun (✓) work for your goals
( ) don't break the rules ( ) goals are too hard

(2) What are the first two rhyming words?
The first two rhyming words are heed and succeed.

(3) Put a check (✓) next to the best title for the poem above.
( ) Success Is the Best
(✓) Try Again
( ) Failing Isn't Fun

DAY 37, pages 73 & 74

❶ 1,000 milliliters (mL) are equal to 1 liter (L). Convert the measurements below.

(1) 1 L = 1,000 mL
(2) 3 L = 3,000 mL
(3) 5 L = 5,000 mL
(4) 1,000 mL = 1 L
(5) 4,000 mL = 4 L
(6) 5,000 mL = 5 L

❷ How much water is in each container below?

(1) (100 mL)
(2) (200 mL)
(3) (400 mL)
(4) (600 mL)
(5) (700 mL)
(6) (900 mL)
(7) (1 L or 1,000 mL)

① Read the poem below, "If I Ever See" by Lydia Maria Child. Then answer the questions.

If ever I see,
On bush or tree,
Young birds in their pretty nest,
I must not in play,
Steal the birds away,
To grieve their mother's breast.
My mother, I know,
Would sorrow so,
Should I be stolen away;
So I'll speak to the birds
In my softest words,
Nor hurt them in my play.
And when they can fly
In the bright blue sky,
They'll warble a song to me;
And then if I'm sad
It will make me glad
To think they are happy and free.

(1) Put a check (✓) next to the phrases that describe the poem's main idea.
( ) catching birds is fun (✓) stealing birds is bad
(✓) wild birds belong in nature ( ) birds are good pets
(✓) wild birds should be free ( ) young birds can't fly

(2) What are the first two rhyming words?
The first two rhyming words are see and tree.

(3) What are the second two rhyming words?
The second pair of rhyming words are play and away.

(4) What will make the speaker in the poem glad?
The speaker in the poem will be glad to think the birds are happy and free.

DAY 38, pages 75 & 76

❶ Write the time under each clock below.

(1) (8:00)
(2) (8:01)
(3) (8:02)
(4) (8:03)
(5) (8:05)
(6) (8:10)

❷ Write the time under each clock below.

(1) (6:05) (2) (7:01) (3) (10:07) (4) (11:10)

① Read the passage from *The Dragonfly of Lookout Mountain* by Judy Hatch. Then answer the questions below.

The new dragonfly nymph dropped into the water. He sank down to the muddy bottom of the pond.

The nymph was an expert hunter. He had six sturdy legs and a ferocious hooked mouthpart which he could shoot out from under his chin. He caught and ate anything which could not get away, and he grew quickly. When he grew too big for his skin, he shed it and formed a new one the next size up.

(1) Complete the chart below.

Characteristics of the dragonfly nymph	
(a) an expert hunter	(c) ferocious hooked mouthpart
(b) six sturdy legs	(d) grows quickly

② Read the passage. Then answer the questions below.

While the spring passed, he grew quickly. He ate insects, then tadpoles, and even small fish. He shed his skin many times. By summer he was two inches long and the largest and most powerful insect in the pond.

A turtle and a trout also lived in his part of the pond. The nymph knew better than to let them see him.

Most of the time he hid in the soft bottom weeds. If they came too close, he swam quickly by squirting water out behind himself, and squeezed into cracks in rocks where he knew they couldn't find him.

(1) Who shed his skin many times?
The nymph shed his skin many times.

(2) What did the nymph eat?
The nymph ate insects, tadpoles, and even small fish.

(3) Where did the nymph hide?
The nymph hid in the soft bottom weeds and in the cracks in rocks.

(4) When was the nymph two inches long?
The nymph was two inches long by summer.

DAY 39, pages 77 & 78

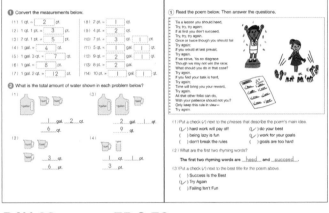

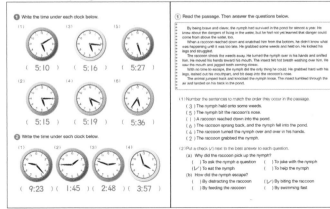

❶ Write the time under each clock below.

(1) (5:10)
(2) (5:15)
(3) (5:16)
(4) (5:19)
(5) (5:27)
(6) (5:36)

❷ Write the time under each clock below.

(1) (9:23) (2) (1:45) (3) (2:48) (4) (3:57)

① Read the passage. Then answer the questions below.

By being brave and clever, the nymph had survived in the pond for almost a year. He knew about the dangers of living in the water, but he had not yet learned that danger could come from above the water, too.

When a raccoon reached down and snatched him from the bottom, he didn't know what was happening until it was too late. He grabbed some weeds and held on. He kicked his legs and struggled.

The raccoon shook the weeds away. He turned the nymph over in his hands and sniffed him. He moved his hands toward his mouth. The insect felt hot breath washing over him. He saw the mouth and jagged teeth coming closer.

With no time to escape, the nymph did the only thing he could. He grabbed hard with his legs, lashed out his mouthpart, and bit deep into the raccoon's nose.

The animal jumped back and knocked the nymph loose. The insect tumbled through the air and landed on his back in the pond.

(1) Number the sentences to match the order they occur in the passage.
(3) The nymph held onto some weeds.
(5) The nymph bit the raccoon's nose.
(1) A raccoon reached down into the pond.
(6) The raccoon sprang back, and the nymph fell into the pond.
(4) The raccoon turned the nymph over and over in his hands.
(2) The raccoon grabbed the nymph.

(2) Put a check (✓) next to the best answer to each question.
(a) Why did the raccoon pick up the nymph?
( ) To ask the nymph a question ( ) To joke with the nymph
(✓) To eat the nymph ( ) To help the nymph
(b) How did the nymph escape?
( ) By distracting the raccoon (✓) By biting the raccoon
( ) By feeding the raccoon ( ) By swimming fast

DAY 40, pages 79 & 80

❶ Sort the shapes into the categories below.

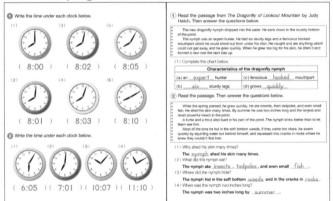

(1) Square (b, d, f)
(2) Rectangle (c, h, i)
(3) Not a square and not a rectangle (a, e, g)

❷ Look at the triangles and list the letters for the right triangles below.

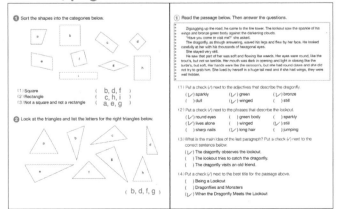

(b, d, f, g)

① Read the passage below. Then answer the questions.

Zigzagging up the road, he came to the fire tower. The lookout saw the sparkle of his wings and bronze green body against the darkening clouds.

"Have you come to visit me?" she asked.

The dragonfly, as though answering, waved his legs and flew by her face. He looked carefully at her with his thousands of hexagonal eyes.

She stayed very still.

He saw that part of her was soft and flowing like weeds. Her eyes were round, like the trout's, but not so terrible. Her mouth was dark in opening and light in closing like the turtle's, but soft. Her hands were like the raccoon's, but she had round claws and she did not try to grab him. She lived by herself in a huge tall nest and if she had wings, they were well hidden.

(1) Put a check (✓) next to the adjectives that describe the dragonfly.
(✓) sparkly (✓) green (✓) bronze
( ) dull (✓) winged ( ) still

(2) Put a check (✓) next to the phrases that describe the lookout.
(✓) round eyes ( ) green body ( ) sparkly
(✓) lives alone ( ) winged (✓) still
( ) sharp nails (✓) long hair ( ) jumping

(3) What is the main idea of the last paragraph? Put a check (✓) next to the correct sentence below.
(✓) The dragonfly observes the lookout.
( ) The lookout tries to catch the dragonfly.
( ) The dragonfly visits an old friend.

(4) Put a check (✓) next to the best title for the passage above.
( ) Being a Lookout
( ) Dragonflies and Monsters
(✓) When the Dragonfly Meets the Lookout

DAY 41, pages 81 & 82

① If you constructed boxes by folding the figures on the left, what kinds of boxes would you make? Connect the figures on the left to the correct boxes on the right.

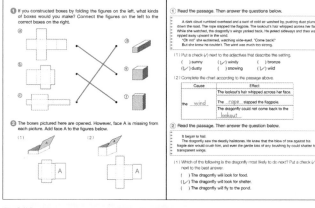

ⓐ
ⓑ
ⓒ

ⓓ
ⓔ
ⓕ

② The boxes pictured here are opened. However, face A is missing from each picture. Add face A to the figures below.

(1)   (2)

① Read the passage. Then answer the questions below.

A dark cloud rumbled overhead and a swirl of cold air washed by, pushing dust plumes down the road. The rope slapped the flagpole. The lookout's hair whipped across her face. While she watched, the dragonfly's wings yanked back. He picked sideways and then was ripped away upward in the wind.
"Oh no!" she exclaimed, watching wide-eyed. "Come back!"
But she knew he couldn't. The wind was much too strong.

(1) Put a check (✓) next to the adjectives that describe the setting.
( ) sunny   (✓) windy   ( ) bronze
(✓) dusty   ( ) snowing   ( ) wild

(2) Complete the chart according to the passage above.

Cause	Effect
the wind	The lookout's hair whipped across her face.
	The rope slapped the flagpole.
	The dragonfly could not come back to the lookout.

② Read the passage. Then answer the question below.

It began to hail.
The dragonfly saw the deadly hailstones. He knew that the blow of one against his fragile skin would crush him, and even the gentle kiss of any brushing by could shatter his transparent wings.

(1) Which of the following is the dragonfly most likely to do next? Put a check (✓) next to the best answer.
( ) The dragonfly will look for food.
(✓) The dragonfly will look for shelter.
( ) The dragonfly will fly to the pond.

DAY 42, pages 83 & 84

① Complete the exercises below based on the charts.

Class A	
Fruit	Number of Students
Apples	10
Oranges	8
Melons	11
Peaches	4
Grapes	2
Pears	3

Class B	
Fruit	Number of Students
Apples	9
Oranges	10
Melons	8
Peaches	5
Grapes	3
Pears	2

Class C	
Fruit	Number of Students
Apples	11
Oranges	7
Melons	10
Peaches	6
Grapes	4
Pears	1

(1) Calculate the total for each type of fruit and write it in the appropriate box in the table on the right.

(2) Which class has the most students?
(class C)

(3) Which fruit is favored by the most students?
(Apple)

(4) What does the number in ⓓ represent?
(Total students)

Class Favorite Fruit

Fruit	Class A	Class B	Class C	Total
Apples	10	9	11	ᵃ 30
Oranges	8	10	7	ᵇ 25
Melons	11	8	10	ᶜ 29
Peaches	4	5	6	15
Grapes	2	3	4	9
Pears	3	2	1	6
Total	ᵃ 38	ᵇ 37	ᶜ 39	ᵈ 114

① Read the passage. Then answer the questions below.

While the morning sun poured strength into the dragonfly, he hung in the tree and watched the meadow come back to life.
New water from the storm overflowed the pond and spilled into the creek. The air hummed with insects. As if with a long sigh of relief, the meadow was recovering after the rain.
The dragonfly watched and realized that this was where he was meant to spend the rest of his life.
He left the pine tree and chose a home. It was a tall cattail by the pond. He defended his territory around the cattail by chasing off other insects when they came too close.

(1) What are the main events of this passage?
(a) After the storm, the dragonfly _hung_ in the tree.
(b) Water _overflowing_ the pond and _spilled_ into the creek.
(c) The air _hummed_ with insects.
(d) The dragonfly _chose_ a _cattail_ as his home.
(e) The dragonfly _defended_ his territory.

(2) Put a check (✓) next to the phrases or words that could be the story's theme.
(✓) home   ( ) friendship   (✓) nature's wonder
(✓) survival   ( ) war   ( ) love

(3) Which of the following is the dragonfly most likely to do next? Put a check (✓) next to the best answer.
(✓) The dragonfly will look for food.
( ) The dragonfly will leave his cattail for vacation.
( ) The dragonfly will choose a new home close to the raccoon.

(4) Put a check (✓) next to the best title for the passage above.
(✓) Dragonfly Finds a Home
( ) The Lookout and the Dragonfly's New Home
( ) Why Cattails Are Good Homes

DAY 43, pages 85 & 86

① Multiply.

(1) $4 \times 2 = 8$
(2) $6 \times 5 = 30$
(3) $8 \times 1 = 8$
(4) $2 \times 6 = 12$
(5) $4 \times 7 = 28$
(6) $8 \times 10 = 80$
(7) $3 \times 3 = 9$
(8) $7 \times 5 = 35$
(9) $9 \times 5 = 45$
(10) $6 \times 7 = 42$
(11) $2 \times 8 = 16$
(12) $3 \times 2 = 6$

② There are 6 children and you want to give them 8 pieces of candy each. How many pieces of candy will you need?

$8 \times 6 = 48$

Ans. 48 pieces of candy

③ Each baseball team is made up of 9 players. If there are 7 teams waiting to play, how many players are there in all?

$9 \times 7 = 63$

Ans. 63 players

④ There are 24 cookies and 7 children. If they divide the cookies equally, how many cookies does each child get? How many cookies are left over?

$24 \div 7 = 3 R 3$

Ans. 3 cookies, 3 cookies left over

① Complete the passage using the vocabulary words defined below.

Did you know that starfish are actually not fish at all? Although most people know them as starfish, many scientists are trying to change the name to sea star. Fish have backbones but sea stars do not. Animals without backbones, like sea stars, are called invertebrates. Animals with backbones, like fish, are called _vertebrates_.
Sea stars are very _unique_ creatures. Sea stars can _regenerate_ a new arm if one is lost. Most have five arms, but some sea stars can grow as many as fifty arms. Their arms have _pincers_, and suckers that help them grip the ground and move.
Most sea stars also have the remarkable ability to _consume_ food outside their bodies. Sea stars use their feet to open clams. The stomach _emerges_ from the mouth and goes inside the clam shell. The stomach then _envelops_ the prey and goes back into the body when it is done.

vertebrates: have a backbone
unique: rare, or one of a kind
pincers: jaws used for gripping or pinching
regenerate: regrow
consume: eat or drink
emerges: moves out or comes into view
envelops: wraps up, covers, or surrounds

② Use the passage above to answer the questions below.

(1) Who wants to change the name of the starfish to sea star?
Scientists want to change the name of the starfish to sea star.
(2) How don't sea stars have?
Sea stars don't have _backbones_.
(3) How do sea stars move?
Sea stars have _pincers_ and _suckers_ that help them _grip_ the ground.
(4) Why don't scientists want to use the name starfish?
Because these animals are not actually _fish_.

DAY 44, pages 87 & 88

① Divide.

(1) $6 \div 2 = 3$
(2) $6 \div 3 = 2$
(3) $8 \div 4 = 2$
(4) $8 \div 3 = 2 R 2$
(5) $15 \div 5 = 3$
(6) $20 \div 4 = 5$
(7) $36 \div 6 = 6$
(8) $42 \div 7 = 6$
(9) $56 \div 8 = 7$
(10) $50 \div 6 = 8 R 2$
(11) $50 \div 7 = 7 R 1$
(12) $63 \div 9 = 7$
(13) $70 \div 8 = 8 R 6$
(14) $48 \div 5 = 9 R 3$
(15) $24 \div 3 = 8$

② We have 24 pencils for my group today. If the 6 of us share them equally, how many pencils will we each get?

$24 \div 6 = 4$

Ans. 4 pencils

③ There are 45 cookies. If 7 children get 6 cookies each, how many cookies are left over?

$45 \div 7 = 6 R 3$

Ans. 3 cookies left over

① Read the passage. Then answer the questions below.

Stephanie and her little brother Nick were walking home from school. Stephanie was blowing into her half-full water bottle and tooting a tune to pass the time. Nick asked, "What makes the sound in your bottle?"
"Well, I could tell you," Stephanie said, "but I think I'll show you too!"
When they got home, Stephanie set up the experiment. She went into the kitchen and got five identical plastic bottles out of the recycling bin. Then she asked Nick to fill them up with different amounts of water. Then she arranged the bottles in order from most full to least full.
"Now blow across the top of each bottle and listen to the sounds," Stephanie instructed.
"They're different!" Nick said after he had blown up and down the row of bottles until he was out of breath.
"That's right! When you blow across the top, you make the air inside shake and shiver. The bottles with more air sound lower than the bottles with more water."

(1) Complete the chart below.

The Bottle Music Experiment
i. Get five _identical_ plastic bottles.
ii. Fill them with different amounts of _water_.
iii. Arrange the bottles from most _full_ to least _full_.
iv. _Blow_ across the top and _listen_ to the sounds.

(2) Complete the chart with words from the passage above.

Cause	Effect
more air in the bottle	The sound is _lower_.
more _water_ in the bottle	The sound is higher.

DAY 45, pages 89 & 90

① Write each fraction.

(1) 1 foot is divided into 5 equal parts, and 2 parts are shaded. $\left(\frac{2}{5}\right)$

(2) 1 foot is divided into 7 equal parts, and 5 parts are shaded. $\left(\frac{5}{7}\right)$

(3) 1 foot is divided into 9 equal parts, and 4 parts are shaded. $\left(\frac{4}{9}\right)$

② Convert the measurements below.

(1) 1 mi. = 5,280 ft.
(2) 1 mi. 3 ft. = 5,283 ft.
(3) 5,290 ft. = 1 mi. 10 ft.
(4) 6,000 ft. = 1 mi. 720 ft.
(5) 1 km 101 m = 1,101 m
(6) 2,010 m = 2 km 10 m

③ Order the following groups from heaviest to lightest with the numbers 1 through 4.

(1) 1 lb.   18 oz.   1 lb. 1 oz.   14 oz.
     (3)   (1)   (2)   (4)

(2) 1,010 g   999 g   1 kg   1 kg 9 g
     (1)   (4)   (3)   (2)

(3) 1,009 mg   1 g   1,100 mg   999 mg
     (2)   (3)   (1)   (4)

④ What is the total amount of water shown in each problem below?

(1)   1 gal. 1 qt.
      5 qt.

(2)   2 gal. 2 qt.
      10 qt.

① Read the passage. Then answer the questions using words from the passage.

The Olympic Games are a sporting event that started in ancient Greece. But it wasn't until 2010 that the Youth Olympic Games was launched. The Youth Olympic Games began at the Singapore Olympics. Its goal is to bring together the world's best young athletes to play and compete in sports, as well as to learn from new cultures. Even non-athletes participate as volunteers, reporters, torch bearers, and more.
For the sports competitions, there are three age groups that can compete: fifteen- and sixteen-year-olds, sixteen- and seventeen-year-olds, and seventeen- and eighteen-year-olds. In the summer, kids play twenty-eight sports—sometimes with boys and girls on the same team. Talented kids from all around the world come together every Olympics, which take place every four years and last for ten to twelve days. In 2010, there were 3,600 athletes. To make sure that kids from every part of the world can play, there are four spots saved for each nation. Some of the sports include pole vaulting, gymnastics, diving, and tennis.

(1) What are the Olympic Games?
The Olympic Games are a _sporting_ _event_ that started in _ancient_ _Greece_.

(2) Who can compete at the Youth Olympic Games?
There are _three_ age groups that can compete: _fifteen- and sixteen-_ year-olds, _sixteen- and seventeen-_ year-olds, and _seventeen-_ _and eighteen-_ year-olds.

(3) Where were the first Youth Olympic Games?
The first Youth Olympic Games were in _Singapore_.

(4) When were the first Youth Olympic Games?
The first Youth Olympic Games were in _2010_.

(5) Why are four spots saved for each nation?
Four spots are saved for each nation to make sure that _kids from_ _every part of the world can play_.

(6) How long does the Youth Olympic Games last?
The Youth Olympic Games last for _ten_ to _twelve_ days.